Cognitive Radar

The Knowledge-Aided Fully Adaptive Approach

Second Edition

For a listing of recent titles in the
Artech House Radar Series,
turn to the back of this book.

Cognitive Radar

The Knowledge-Aided Fully Adaptive Approach

Second Edition

Joseph R. Guerci

ARTECH HOUSE
BOSTON | LONDON
artechhouse.com

Library of Congress Cataloging-in-Publication Data
A catalog record for this book is available from the U.S. Library of Congress.

British Library Cataloguing in Publication Data
A catalog record for this book is available from the British Library.

ISBN-13: 978-1-63081-773-2

Cover design by John Gomes

© 2020 Artech House

All rights reserved. Printed and bound in the United States of America. No part of this book may be reproduced or utilized in any form or by any means, electronic or mechanical, including photocopying, recording, or by any information storage and retrieval system, without permission in writing from the publisher.

All terms mentioned in this book that are known to be trademarks or service marks have been appropriately capitalized. Artech House cannot attest to the accuracy of this information. Use of a term in this book should not be regarded as affecting the validity of any trademark or service mark.

10 9 8 7 6 5 4 3 2 1

Contents

Preface   11

1

Introduction   15

1.1 Why "Cognitive" Radar?   15
1.2 Functional Elements and Characteristics of a Cognitive Radar Architecture   16
1.2.1 Adaptive Transmit Capability   19
1.2.2 Knowledge-Aided Processing   24
1.2.3 Optimum Resource Allocation and Scheduling for Cognitive Radar   29
1.2.3 Optimum Resource Allocation and Scheduling for Cognitive Radar   31
1.3 Book Organization   32
References   32

2

Optimum Multi-Input Multioutput Radar   37

2.1 Introduction   37
2.2 Jointly Optimizing the Transmit and

Receive Functions
Case I: Maximizing SINR   38
Example 2.1 Multipath Interference   43
2.3 Jointly Optimizing the Transmit and Receive Functions
Case II: Maximizing Signal-to-Clutter   48
Example 2.2 Sidelobe Target Suppression: Sidelobe Nulling on Transmit   49
Example 2.3 Optimal Pulse Shape for Maximizing SCR   51
Example 2.4 Optimum Space-Time MIMO Processing for
Clutter Suppression in Airborne MTI Radar   54
2.4 Optimum MIMO Target Identification   60
Example 2.5 Two-Target Identification Example   63
Multitarget Case   64
Example 2.6 Multitarget Identification Example   66
2.5 Constrained Optimum MIMO Radar   67
Case I: Linear Constraints   67
Example 2.7 Prenulling on Transmit   68
Case II: Nonlinear Constraints   70
Relaxed Projection Approach   70
Example 2.8 Relaxed Projection Example   71
Constant Modulus and the Method of Stationary Phase   72
Example 2.9 Nonlinear FM (NLFM) to Achieve Constant Modulus   74
Example 2.10 Matched Subspace Example   80
2.6 Recent Advances in Constrained Optimum MIMO   80

References   82

Appendix 2.A: Infinite Duration (Steady State) Case 84

3

Adaptive MIMO Radar 87

3.1 Introduction 87

3.2 Transmit-Independent Channel Estimation 88

Example 3.1 Adaptive Multipath Interference Mitigation 90

3.3 Dynamic MIMO Calibration 91

Example 3.2 MIMO Cohere-on-Target 91

3.4 Transmit-Dependent Channel Estimation 93

Example 3.2 STAP-Tx Example 94

Example 3.2 DDMA MIMO STAP Clutter Mitigation Example for GMTI Radar 97

3.5 Theoretical Performance Bounds of the DDMA MIMO STAP Approach 101

3.6 Nonorthogonal MIMO Probing for Channel Estimation 106

References 116

4

Introduction to KA Adaptive Radar 119

4.1 The Need for KA Radar 119

4.2 Introduction to KA Radar: Back to "Bayes-ics" 124

4.2.1 Indirect KA Radar: Intelligent Training and Filter Selection 126

Example 4.1 Intelligent Filter Selection: Matching the Adaptive DoFs (ADoFs) to the Available Training Data 128

4.2.2 Direct KA Radar: Bayesian Filtering and Data Prewhitening 131
Example 4.2 Using Past Observations as a Prior Knowledge Source 135
4.3 Real-Time KA Radar: The DARPA KASSPER Project 139
4.3.1 Solution: Look-Ahead Scheduling 140
Example 4.3 Balancing Throughput in a KASSPER HPEC Architecture 143
4.3.2 Examples of a KA Architectures Developed by the DARPA/AFRL KASSPER Project 146
4.4 KA Radar Epilogue 153
References 155

5

Putting it All Together: CoFAR 159

5.1 Cognitive Radar: The Fully Adaptive Knowledge-Aided Approach 159
5.1.1 A Cognitive Radar Architecture for GMTI 160
5.1.2 Informal Operational Narrative for a GMTI Radar 162
5.2 CoFAR Radar Scheduler 164
5.3 Areas for Future Research and Development 166
References 167

6

Cognitive Radar and Artificial Intelligence 169

6.1 Relationship between Cognitive Radar and Artificial Intelligence 169

6.2 Cognitive Radar Utilizing Traditional AI   170
6.3 Cognitive Radar Utilizing Deep Learning AI   171
6.3.1 CoFAR Mission Computer   173
6.3.2 CoFAR Radar Controller and Scheduler   174
6.3.3 CoFAR RTCE   175
6.4 Summary   175

References   176

About the Author   177

Index   179

Preface

It has been 10 years since the publication of the first edition of *Cognitive Radar: The Knowledge-Aided Fully Adaptive Approach,* and the popularity and interest in cognitive radar (CR) has increased dramatically. Since 2010, there have been well over 100 peer-reviewed publications specifically on cognitive radar (source: Google Scholar), not including closely related topics such as waveform adaptivity and joint radar and communications coexistence [1]. Though originally developed for demanding military environments, the first successful commercial cognitive radar for autonomous vehicles was recently announced (see www.echodyne.com). Clearly, CR is here to stay.

Unfortunately, there is still a lot of confusion and debate over what exactly a CR is. A major source of confusion is an erroneous use of terms such as consciousness, machine learning, artificial intelligence (AI), and deep learning as *synonyms* for CR—which is incorrect. As described in Chapter 1, and in the original first edition, the term cognition is a very well defined scientific and engineering term that pertains to any system human or otherwise, that has contextual sensory awareness of its environment, can solve problems, and retains memories (see Chapter 1 for a precise definition from the National Institute of Health, along with its direct translation into engineering terms).

Of course, a CR can and should leverage any and all tools from AI that can improve its adaptation—particularly in highly complex and contested environments. For example, deep learning networks have more than proven their worth at advanced pattern recognition for advanced target ID against a cluttered background. Chapter 6 in this second edition discusses the use of advanced AI methods in CR. However, and this is crucial, CR does not require so-called modern AI, which has become synonymous with deep learning. For example, one of the most crucial functions of a CR is real-time adaptive channel estimation, which is estimation of the entire radar channel that in general consists of targets, clutter, and interference (intentional or otherwise). As shown in Chapters 2 and 3, there are optimum probing techniques that can greatly enhance channel estimation that do not, in and of themselves, require AI techniques. That is not to say that AI techniques do not apply, just that they are not a fundamental requirement. Importantly, it is wise to remember always: *as goes channel knowledge, so goes performance!* It follows that: *no channel knowledge, no CR!* As a consequence, this second edition contains new material on the latest sense-learn-adapt (SLA) channel probing techniques leveraging recent advances in multi-input, multioutput (MIMO) radar to further enhance channel knowledge [2].

A new discussion is included in Chapter 5 on CR schedulers. Arguably, this is where the rubber hits the road with cognitive radar. The CR scheduler must not only optimize the real-time allocation of radar resources (timeline, energy, processing, etc.), but it must also utilize resources to enhance channel knowledge for signal-dependent channel elements such as clutter, targets, and potentially responsive/adaptive jamming.

Lastly, in Chapter 6, new material on the relationship between CR and traditional and emerging AI is discussed. Functions of a CR that *could* leverage modern AI are highlighted. However, despite the recent successes of deep learning, real-world radar operations in challenging environments still (at the time of this writing) represent a formidable problem.

The author is greatly indebted to many colleagues with whom he has derived great benefit in developing CR over the past 20+ years. In particular, he wishes to acknowledge (alphabetical) Amy Alving, Augusto Aubry, Chris Baker, Ed Baranoski, Charles Baylis, Kristine Bell, Jamie Bergin, Dan Bliss, Shannon Blunt, Matt Brandsema, Jim Carlini, Vasu Chakravarthy, Tim Clark, Antonio DeMaio, Tom Driscoll, Yonina Eldar, Alfonso Farina, Lou Fertig, Fulvio Gini, Sandeep Gogineni, Scott Goldstein, Nathan Goodman, Martie Goulding, Maria Greco, Marshall Greenspan, Hugh Griffiths, Evelyn Guerci, Ray Guerci, Simon Haykin, Braham Himed, Dave Kirk, Alex Lackpour, Jian LiLorenzo LoMonte, Anthony Martone, William Melvin, Ram Narayanan, Arye Nehorai, Hoan Nguyen, Aram Partizian, Mike Picciolo, Unnikrishna Pillai, Muralidhar Rangaswamy, Joe Schuster, Greg Showman, Graeme Smith, Dan Stevens, Dan Stevens, Petre Stoica, Nik Subotic, Paul Techau, Tony Tether, Brian Watson, Michael Wicks, David Zasada, Michael Zatman, Peter Zulch, and many others!

References

[1] Lackpour, A., A. Rosenwinkel, J. R. Guerci, A. Mody, and D. Ryan, "Design and Analysis of an Information Exchange-Based Radar/Communications Spectrum Sharing System (RCS3)," in *2016 IEEE Radar Conference (RadarConf)*, 2016, pp. 1–6.

[2] Bergin, J. S., and J. R. Guerci, *Introduction to MIMO Radar*, Norwood, MA: Artech House, 2018.

1

Introduction

1.1 Why "Cognitive" Radar?

There are, not surprisingly, many variations of the definition of cognition. However, a succinct yet comprehensive description that will be of relevance to cognitive radar as described herein is afforded by the National Institutes of Health (NIH), National Institute of Mental Health (NIMH) [1]:

> Cognition: Conscious mental activity that informs a person about his or her environment. Cognitive actions include perceiving, thinking, reasoning, judging, problem solving, and remembering.

However these faculties arose in humans, there is no disputing their value to survival despite the considerable physiological cost and accommodations that are required to possess them. Simply stated, being "smart" when it comes to environmental interactions in all its varied forms is well worth it. And while true thinking machines are still very much the stuff of science fiction, it is still possible to be guided by the above principles and map each of the above cognitive properties into real engineering systems. Just such a mapping that is the basis of this book is provided in Table 1.1 and is distributed in detail throughout Chapters 2 to 6.

So what exactly are the potential benefits of a radar possessing some manner of the aforementioned cognitive abilities? This of course depends on the type of radar, its mission, and the environment in which it must operate. One of the most formidable is that of a modern ground moving target indicator (GMTI) radar that attempts to discern "moving" man-made ground-based targets from a multitude of competing environmental phenomena. Indeed it is the author's own experience in this regime that was the impetus for many of the developments described in this book. Table 1.2 lists a few of the complex real-world environmental effects that confront a modern GMTI radar—all of which have been shown to, or in principle should, significantly benefit from cognitive radar (CR) abilities [2].

1.2 Functional Elements and Characteristics of a Cognitive Radar Architecture

In this section we will introduce and overview the salient functional elements and characteristics of a proposed cognitive radar. To that end, consider the basic radar block diagrams of Figure 1.1, which depict a conventional adaptive radar (a), and that of a cognitive fully adaptive radar (CoFAR) (b). Though a conventional radar certainly is adaptive, it is usually confined to the receiver

Table 1.1
Mapping of Biological Cognitive Properties to a Cognitive Radar

Cognitive Property	Cognitive Radar Equivalent
Perceiving	Sensing
Thinking, reasoning, judging, problem solving	Expert systems, rule-based reasoning, adaptive algorithms, artificial intelligence, and computation
Remembering	Memory, environmental database, and knowledge-aided (KA) processing (see Chapter 4)

1.2 Functional Elements and Characteristics of a Cognitive Radar Architecture 17

Table 1.2
Examples of Real-World Phenomenon and Their Potential Deleterious
Impacts on Performance That in Principle Can Be Alleviated by
Incorporating Cognitive Radar Concepts

Phenomenon	Illustration	Impact
Heterogeneous clutter		Over/under-nulled clutter, increased false alarms, desensitization, lack of training data
Dense-target backgrounds		Desensitization, lack of training data, increased false alarms
Large discretes, large man-made structures		Increased false alarms, target masking
Configuration-induced nonstationary clutter behavior (e.g., bistatic radar, and nonlinear arrays)		Lack of training data
Electronic countermeasures		Varying impact

and is often very reactive to the received data stream—that is, adaptivity is based solely on the very data stream to be interrogated for targets [3–5]. There is very little provision for learning over time, feedback to the transmitter, or the integration of exogenous environmental information sources that can provide significant benefits, such as digital terrain maps.

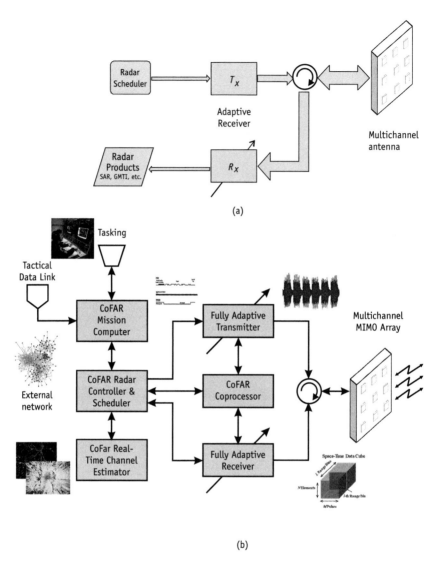

Figure 1.1 (a) High-level block diagrams of a conventional modern radar, and (b) the new CoFAR architecture.

In contrast, the cognitive radar exhibits a number of advanced elements that could be argued to better mimic biologically cognitive systems:

- An environmental dynamic database (EDDB) that contains knowledge of the environment and/or targets of interest gleaned from both onboard (endogenous) and off-board (exogenous) information sources [6]. The EDDB is a primary component of knowledge-aided (KA) methods such as KA-STAP.
- In addition to an adaptive receiver, the cognitive radar includes an adaptive transmitter—and thus provision for feedback from the receive chain. Indeed the presence of this feedback has been identified by Haykin [7] as an essential ingredient in any cognitive radar. The multiple transmit degrees-of-freedom (DoFs) also gives rise to the concept of a multi-input (transmitter), multioutput (receiver), or MIMO radar [8–10].
- Though not explicitly shown, the receiver and transmitter adaptivity entail a number of new adaptation and KA methods that are described in Chapters 2 to 4 (and references cited therein).
- A CR resource optimization and scheduling function that makes intelligent decisions on-the-fly

In the next sections, we will briefly overview both the adaptive transmitter and knowledge-aided methods that are the major new elements being introduced into adaptive radar.

1.2.1 Adaptive Transmit Capability

Until recently, modern adaptive radar was synonymous with receiver adaptivity [11]. While most radars have some form of primitive transmit adaptivity, usually in the form of mode selection (e.g., track versus search or, low versus high resolution), true adaptivity involving the continuous signal-dependent variation of one or more of the transmit DoFs is only recently emerging in the radar theoretic literature [12–16]. There are good reasons for this lag: Adaptivity on transmit generally entails advanced front-end hardware (e.g., digital arbitrary waveform generators (DAWGS) and solid-state transmitters), and the availability of new transmit adaptation algorithms. With the advent of the next generation of digital radars [17], the first technical excuse has been removed.

Thus, assuming the hardware exists to support transmit adaptivity, what are the potential benefits? This is the subject of Chapter 2, which derives, from first principles, optimum transmitter designs for a multitude of applications and DoFs. While the reader is referred to Chapter 2 (and references cited therein) for details, we will sample a few of the potential benefits of transmit adaptivity to motivate this new radar functionality.

Figure 1.2 shows the benefits of tailoring the transmit waveform (fast-time modulation) to account for a colored noise RF interference source. Note that optimum waveform design techniques can be used to optimally redistribute the transmit RF spectrum to maximize the signal-to-interference-plus-noise ratio (SINR). Such a waveform tailoring technique can be used to minimize the impact of cochannel interference, as well as minimize the radar's impact on other cochannel receivers (e.g., transmit spectrum shaping). The inclusion of other waveform constraints (such as constant modulus) are discussed in Chapter 2 and the references cited therein.

The above example illustrates the optimization of but one of the transmit DoFs: fast-time (i.e., waveform) modulation. Other DoFs potentially available for adaptive optimization include spatial (azimuth/elevation antenna patterns), polarization, and possibly slow-time modulation (e.g., Doppler division multiple access (DDMA) MIMO radar [8, 9, 18]). Figure 1.3 shows how the transmit pattern can be optimized to minimize the impact of strong sidelobe targets (and/or clutter discretes). By proactively attenuating the signals on transmit, the problem of sidelobe target leakage [19] can be effectively ameliorated.

Interestingly, optimization of the transmit DoFs can be extended to the problem of target identification—indeed it is one of the earliest applications of transmit diversity in the form of matched illumination [12–16]. Figure 1.4(a) shows the baseband impulse responses for two different targets. The spectrum of the optimum waveform that maximally separates the responses for the two hypothesized targets is shown in Figure 1.4(b), along with a conventional chirp. The distance metric, d, as discussed in Chapter 2, measures the degree of separation in measurement space between the two responses. The larger d, the easier it is to ascertain which

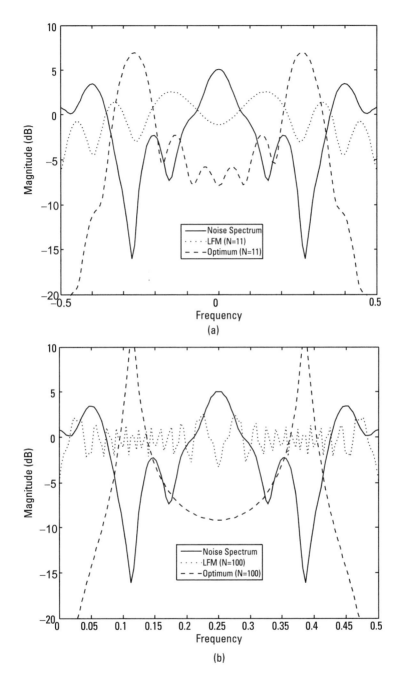

Figure 1.2 (a, b) Example of the benefits of using optimum transmit waveform shaping to maximize SINR in the presence of cochannel RF interference (see Example 2.1, Chapter 2). Note how the optimum pulse is antimatched to the colored noise spectrum, in contrast to the nonoptimized LFM waveform. (a) Short pulse length, and (b) long pulse length.

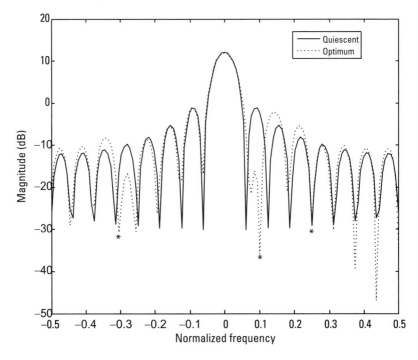

Figure 1.3 Example of tailoring the spatial transmit pattern to proactively ameliorate the problem of sidelobe target leakage (see Example 2.3, Chapter 2). Note the presence of transmit nulls in the directions of the sidelobe targets.

target is likely present (assuming adequate SNR for detection). For this example, there is approximately a 7-dB gain in separability between the optimum and nonoptimized (chirp) probing pulses.

The idea of adapting the illumination in response to the channel runs deep, though it is only recently that radar technology has caught up. As far as this author has been able to discover, the earliest reference to "optimizing the input signal for a known channel" in order to maximize the output response can be traced to a 1950 paper by Chalk [20], which was for the single-input, single-output (SISO), real-valued signal case. Manasse addressed the optimal design of a waveform in the presence of homogenous distributed clutter [21], followed by the treatment of Van Trees who similarly addressed the waveform design problem for reverberation channels (i.e., clutter) in his seminal multivolume treatise [22].

In the late 1980s and early 1990s there was a resurgence of interest in waveform adaptation to maximize the response of

1.2 Functional Elements and Characteristics of a Cognitive Radar Architecture

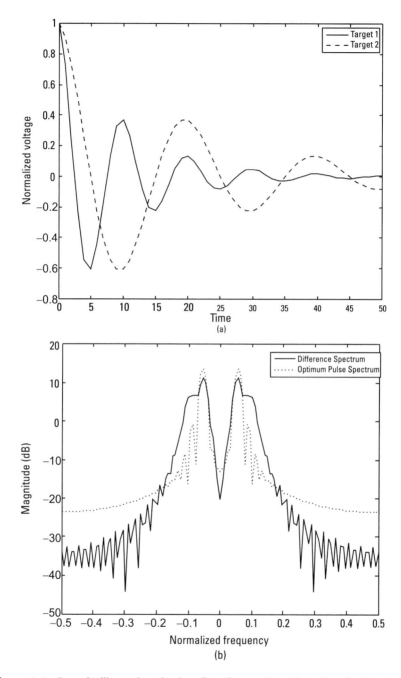

Figure 1.4 Example illustrating the benefits of transmit optimization for the target identification problem. (a) Baseband impulse responses for two different targets. (b) Spectrum of optimum pulse illustrating its matching to the two-target difference spectrum resulting in improved target ID (see Chapter 2, Example 2.5, for further details).

(presumably) weak targets. See references [12–16] for early examples of matched illumination for both maximizing signal response, as well as enhancing the target identification problem. The treatment in this book (Chapters 2 and 3) are based on the above, along with the author's own research over a number of years beginning in the late 1980s through to the present.

Both an adaptive receiver and transmitter require useful knowledge of the channel (interference and targets). Traditional adaptation paradigms extract information solely from the observed data— usually during a so-called "training" period that is integrated into the embedded processing and rapidly performed to maintain requisite throughput requirements.

In the next section we discuss a new approach to adaptation that achieves the remembering function of cognition described previously, and allows for both endogenous and exogenous information sources.

1.2.2 Knowledge-Aided Processing

While adaptivity that allows for transmit modification based on observed environmental/target signals is, in and of itself, an essential ingredient in cognition, (Chapters 2 and 3 are devoted to this topic), its impact can be greatly enhanced if even more effective means of environmental awareness are employed.

Knowledge-aided (KA) processing as developed under the DARPA/AFRL KASSPER project (knowledge-aided sensor signal processing and expert reasoning), essentially consists of an environmental dynamic database populated in general by both endogenous and exogenous information sources, and an embedded computing architecture that integrates the EDDB into the real-time radar signal processor [6] (see Figure 1.5). While it is somewhat intuitive that onboard sensor observation histories (i.e., endogenous sources) provide an indispensable source of environmental awareness, it should with modest reflection be equally obvious that many exogenous (off-board) sources can likewise be quite valuable. For example, through the use of geospatial databases it is possible to precisely know where various geographical features impacting radar performance are located. These include, but are not limited to, road networks, terrain elevation/

1.2 Functional Elements and Characteristics of a Cognitive Radar Architecture 25

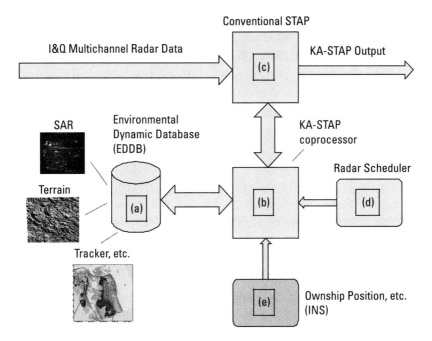

Figure 1.5 Illustration of a canonical KA-STAP architecture featuring an EDDB, labeled (a), a KA coprocessor (b) that works in concert with the conventional (i.e., causal) STAP unit (c) and performs the look-ahead (noncausal) function that is made possible by the predictive nature of the radar scheduler (d) and ownship position/velocity information unit (e).

slope and coverage types, as well as the location of potentially strong discrete clutter returns (e.g., buildings, towers, power lines, and land-sea interfaces) [2]. Though intuitively obvious that such information is of potential benefit, it is quite another matter as to precisely how to utilize it in a real-time radar signal processor—which is the subject matter of Chapter 4.

Though the reader is referred to Chapter 4 for further details associated with the KA-STAP architecture, its basic operation is relatively easy to describe. In normal operation, the KA coprocessor interrogates the EDDB in regions the radar will soon be illuminating to determine if any special KA processing is required. Special KA processing is generally indicated when confronted with target/environmental effects that are not easily managed using traditional methods that rely on stationary/homogenous statistical behavior. The look-ahead function is accomplished by querying the radar scheduler (typically just a few seconds at most in the future), and predicting where the radar platform will be located at this future time—which is readily accomplished by extrapolating position, velocity, acceleration, and so forth, obtained from the ownship inertial navigation system (INS) (as indicated in Figure 1.5). Some examples of situations in which KA processing can significantly improve performance compared to a conventional STAP processor that inherently assumes environmental homogeneity are depicted in Figure 1.6. Again the reader is referred to Chapter 4 for further details.

The origins of radar KA processing as described herein can be traced to the pioneering work conducted at the Air Force's Research Laboratory (AFRL) at Rome, New York. Beginning with expert systems CFAR (constant false alarm rate) [23–26], followed in later years by knowledge-based space-time adaptive processing (KB-STAP) [19, 27–31].

An expert system, from an engineering perspective, attempts to capture the experience and judgment of a human expert in a suitably codified engineered system/algorithm such that to an outside observer the system response to external stimuli mimics that of an expert. This is generally accomplished via some combination of rule-based reasoning [32] and/or adaptive algorithmic structures. The idea behind expert systems CFAR was to improve upon the sophistication of traditional CFAR techniques that

- Heterogeneous Clutter/Multipath
  - Rapidly varying terrain
    - Mountainous (rapid elevation/reflectivity variation)
    - Rapid land cover variations (e.g., littoral)
- Dense "Target" Backgrounds
  - "Moving Clutter"
    - Military/civilian vehicles
- Large Discretes and "Spiky" Clutter
  - Urban clutter
  - Power lines, towers, steep mountainous terrain
- Range-Varying (Nonstationary) Clutter Loci
  - Bi/Multistatics
  - Nonlinear array geometries (e.g., circular arrays)

Figure 1.6 Examples of situations where KA processing can significantly improve performance compared to a conventional STAP approach that inherently assumes environmental homogeneity (see Chapter 4 for further details).

typically used some cell averaging approach (mean, median, etc. [33]) based solely on the observed sensor data without the benefit of any exogenous environmental awareness.

KB-STAP extended the above ideas to the multidimensional filtering problem (CFAR is typically a scalar problem). Traditional STAP is at its core a sample covariance-based technique—and thus in essence equivalent to the aforementioned traditional CFAR. KB-STAP research greatly expanded the potential information sources utilized in the overall adaptation process. An early but compelling example of the benefits of adopting a KB-STAP approach was developed by Melvin et al. [30]. By using knowledge of where the road networks (sources of moving clutter) were located in relation to the radar, and a suitable intelligent training strategy (see Chapter 4), the velocity desensitization problem was circumvented [30].

Beginning in 2001, the Defense Advanced Research Projects Agency (DARPA) and AFRL formed a partnership to pursue the aforementioned KASSPER project—the goal being the development of an entirely new real-time KA-embedded computing architecture that can accommodate the types of KB/KA algorithms that were being developed [2]. While evidence was mounting demonstrating the benefits of enhanced environmental knowledge gleaned from both endogenous and exogenous sources, a fundamental obstacle remained: "The 'knowledge' contained in the EDDB is memory based, and thus in general subject to access latencies that can all but preclude its integration into a modern pipelined embedded computing architecture."

For example, a STAP beamformer is typically based on some variant of a Gram-Schmit or QR-factorization formulation that is amenable to a parallel processing approach—and thus in turn appropriate for real-time processing [34, 35]. The aforementioned knowledge, being memory based, cannot be integrated into the STAP processor without incurring an intolerable latency (see Figure 1.7 for a relative comparison of memory access latencies). When confronted with such a seemingly insurmountable obstacle, one is always advised to reexamine each and every "unquestioned" assumption, and to take a broader "systems" perspective for possible relief—precisely the approach adopted by the DARPA/AFRL KASSPER project.

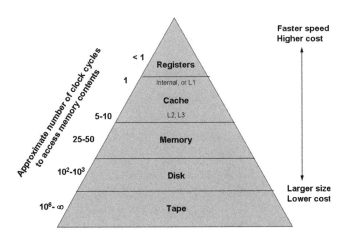

Figure 1.7 One of the keys to the success of the DARPA/AFRL KASSPER project was the development of look-ahead techniques that overcome the significant memory access latency problem in high-performance-embedded computing (HPEC) (see Chapter 4 for further details).

Figure 1.8 illustrates the key to unlocking the real-time KA problem. In essence it is predicated on being able to predict future events a few seconds (at most) in advance. It is thus possible to have a noncausal processor, working in tandem with the causal (traditional) STAP processor, that can perform look-ahead sufficient to compensate for the aforementioned memory access latency (see Figure 1.5). Is this really possible? Consider the following facts:

- If the position of the radar platform is known with good precision at, say $t = 0$, then it is eminently reasonable to expect that a prediction of its future position, say one or two seconds in the future, is of almost equal quality. This is, for example, readily accomplished through the use of a Kalman filter (predictor to be more precise) that uses its plant model and any known control inputs [36].

- Every modern radar possesses a scheduler that tells the radar what to do and when to do it. This radar-operating plan in practice is essentially deterministic, at least for a few seconds into the future (often much longer).

1.2 Functional Elements and Characteristics of a Cognitive Radar Architecture

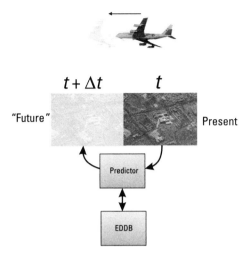

Figure 1.8 Illustration of the ability to accurately predict, a few seconds into the future, where the radar will be and what it will be doing. This is one of the key elements of overcoming the aforementioned memory access latency problem (see Chapter 4 for further details).

If the radar knows, for all intents and purposes, where it will be and what it will be doing (e.g., frequency, waveform, and look-angle) far enough in advance to more than compensate for the memory access latency, then why not perform the necessary knowledge-based calculations in advance of the arrival of the anticipated data? While perhaps insultingly obvious, this nonetheless was the key breakthrough that has allowed for the actualization of KA/KB performance gains in real-time systems.

Figures 1.9 to 1.10 demonstrate the significant performance gains that have been achieved using two different KA-STAP architectures; One developed by Information Systems Laboratories, Inc., (IKA-STAP) [37], the other by Georgia Tech Research Institute (GKA-STAP) [38] (see Chapter 4 for further details).

1.2.3 Optimum Resource Allocation and Scheduling for Cognitive Radar

An extremely important function for any cognitive radar is real-time decision making about radar resource allocation and scheduling. That is because for an advanced CR operating in a complex environment, there will likely be many competing requirements vying for radar resources and timeline. For example, in a target-

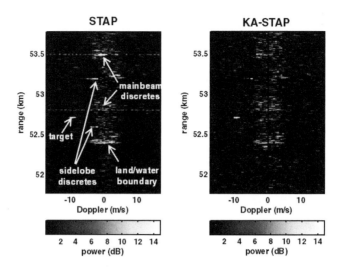

Figure 1.9 Application of Information Systems Laboratories (ISL) knowledge-aided STAP (IKA-STAP) architecture to the 6th KASSPER challenge data set. Conventional STAP (left figure) results in significant clutter leakage and sidelobe discrete targets, while IKA-STAP essentially eliminates all of the complex clutter issues (see Chapter 4 for further details).

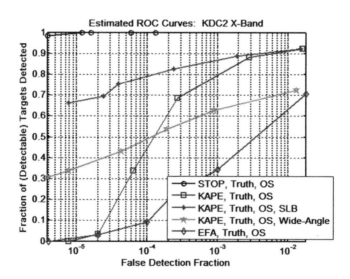

Figure 1.10 Estimated receiver operating characteristic (ROC) curves for the KASSPER challenge data set #2 for the clairvoyant (unobtainable) optimum (STOP) and several GTRI KA variants (prefix KAPE), as compared with a conventional post-Doppler STAP approach. Note that all of the KA methods significantly outperform conventional STAP (see Chapter 4 for further details).

rich environment, there can quickly arise a conflict between the requirement to optimize search area coverage rate (ACR) while simultaneously maintaining a high-quality track on high-value targets (HVTs). In wide area search (WAS) mode, transmit beams need to scan an entire field-of-regard (FOR), dwelling in each beam position just long enough to achieve a requisite detection probability for a prescribed false alarm rate. However, in HVT track mode, it is necessary to dwell longer on a target and/or more frequently revisit the target to maintain a high-quality track. Moreover, additional competing functions, such as target ID and high-resolution SAR imaging, can also arise. It is the job of the CR radar scheduler to optimize the allocation of resources and timeline to satisfy all requirements to the extent possible. While radar schedulers certainly exist in many conventional radars, CR scheduler incorporate far more contextual awareness and the ability to learn via the aforementioned sense-learn-adapt (SLA) perception cycle. See Chapter 5 for further details.

1.2.3 Optimum Resource Allocation and Scheduling for Cognitive Radar

An extremely important function for any cognitive radar is real-time decision making about radar resource allocation and scheduling [39–43]. That is because for an advanced CR operating in a complex environment, there will likely be many competing requirements vying for radar resources and timeline. For example, in a target-rich environment, there can quickly arise a conflict between the requirement to optimize search area coverage rate (ACR), while simultaneously maintaining a high-quality track on high-value targets (HVTs). In wide area search (WAS) mode, transmit beams need to scan an entire field-of-regard (FOR), dwelling in in each beam position just long enough to achieve a requisite detection probability for a prescribed false alarm rate. However, in HVT track mode, it is necessary to dwell longer on a target, and/or more frequently revisit the target to maintain a high-quality track. Moreover, additional competing functions, such as target ID and high-resolution SAR imaging, can also arise. It is the job of the CR radar scheduler to optimize the allocation of resources and timeline to satisfy all requirements to the extent possible. While radar schedulers certainly exist in many conventional radars, CR schedulers incorporate far more contextual awareness and the ability to learn via the aforementioned sense-learn-adapt (SLA) perception cycle. See Chapter 5 for further details.

1.3 Book Organization

Chapter 2 introduces optimum, and constrained optimum, multidimensional transmit signal design (waveform, spatial, etc.) for the purposes of maximizing SINR, as well as maximizing signal separability for the target identification problem. Chapter 3 extends these concepts to the adaptive case (simultaneous channel estimation and signal design), and develops a connection between optimal and orthogonal MIMO radar techniques. Chapters 2 and 3 are essentially a self-contained introduction to optimum and adaptive multi-input, multioutput (MIMO) radar.

Chapter 4 introduces KA methods, including both indirect techniques such as intelligent training and filter selection (ITFS), as well as the more direct (and thus more complex) KA methods such as Bayesian covariance blending. Salient elements of a real-time KA-STAP architecture are then introduced and illustrated via two real-world systems developed by Georgia Tech Research Institute and Information Systems Laboratories, Inc.

Lastly, in Chapter 5, adaptive transmit-receive (i.e., MIMO) is combined with KA to yield the cognitive radar architecture depicted in Figure 1.1. The chapter concludes with a discussion of current areas of development and research.

In Chapter 5, a day in the life of a CR applied to GMTI radar is discussed to help illustrate exactly how CR functions. A new discussion of CR scheduling has also been added.

Lastly, in Chapter 6, an updated discussion is provided on the relationship between CR and modern AI techniques including deep learning.

References

[1] National Institutes of Health (NIH), National Institute of Mental Health (NIMH), pp. "Definition of Cognition,"<http://science-education.nih.gov/supplements/nih5/Mental/other/glossary.htm>.

[2] Guerci, J. R., and W. Baldygo, *Proceedings of the DARPA/AFRL Knowledge-Aided Sensor Signal Processing and Expert Reasoning (KASSPER) Workshop,* 2002–2006.

[3] Guerci, J. R., *Space-Time Adaptive Processing for Radar,* Norwood, MA: Artech House, 2003.

[4] Klemm, R., *Principles of Space-Time Adaptive Processing,* Vol. 12, Institution of Electrical Engineers, 2002.

[5] Ward, J., "Space-Time Adaptive Processing for Airborne Radar," *IEE Colloquium on Space-Time Adaptive Processing,* 1998, p. 2.

[6] Guerci, J. R., and E. J. Baranoski, "Knowledge-Aided Adaptive Radar at DARPA: An Overview," *IEEE Signal Processing Magazine,* Vol. 23, 2006, pp. 41–50.

[7] Haykin, S., "Cognitive Radar: A Way of the Future," *IEEE Signal Processing Magazine,* Vol. 23, 2006, pp. 30–40.

[8] Bliss, D. W., and K. W. Forsythe, "Multiple-Input Multiple-Output (MIMO) Radar and Imaging: Degrees of Freedom and Resolution," *Asilomar Conference on Signals, Systems and Computers Conference Record of the Thirty-Seventh* Vol. 1, 2003, pp. 54–59.

[9] Bliss, D. W., et al., "GMTI MIMO Radar," *Waveform Diversity and Design Conference, International,* 2009, pp. 118–122.

[10] Guerci, J. R., et al., "Theory and Application of Optimum and Adaptive MIMO Radar," *Radar Conference, RADAR '08 IEEE,* 2008, pp.1–6.

[11] Brennan, L., and L. Reed, "Theory of Adaptive Radar," *IEEE Transactions on Aerospace and Electronic Systems,* 1973, pp. 237–252.

[12] Farina, A., and F. A. Studer, "Detection with High Resolution Radar: Great Promise, Big Challenge," *Microwave Journal,* May 1991.

[13] Gjessing, D., *Target Adaptive Matched Illumination Radar: Principles & Applications,* Peter Peregrinus Ltd, 1986.

[14] Grieve, P. G., and J. R. Guerci, "Optimum Matched Illumination-Reception Radar," U.S. Patent 5,175,552, 1992.

[15] Guerci, J. R., "Optimum Matched Illumination-Reception Radar for Target Classification," U.S. Patent 5,381,154, 1995.

[16] Guerci, J. R., and P. G. Grieve, "Optimum Matched Illumination Waveform Design Process," U.S. Patent 5,121,125, 1992.

[17] Szu, H., R. Stapleton, and F. Willwerth, "Digital Radar Commercial Applications," *The Record of the IEEE 2000 International Radar Conference,* 2000, pp. 717–722.

[18] Mecca, V., J. Krolik, and F. Robey, "Beamspace Slow-Time MIMO Radar for Multipath Clutter Mitigation," *IEEE International Conference on Acoustics, Speech and Signal Processing*, ICASSP, 2008, pp. 2313–2316.

[19] Adve, R., T. Hale, and M. Wicks, "Knowledge-Based Adaptive Processing for Ground Moving Target Indication," *Digital Signal Processing*, Vol. 17, 2007, pp. 495–514.

[20] Chalk, J. H. H., "The Optimum Pulse Shape for Pulse Communication," *Proceedings of the Institute of Electrical Engineering*, Vol. 87, 1950, pp. 88–92.

[21] Manasse, R., "The Use of Pulse Coding to Discriminate Against Clutter," *Defense Technical Information Center (DTIC)*, Vol. AD0260230, June 7, 1961.

[22] van Trees, H. L., *Detection, Estimation, and Modulation Theory: Radar-Sonar Signal Processing and Gaussian Signals in Noise*, Krieger Publishing Co., Inc., 1992.

[23] Baldygo, W., et al., "Artificial Intelligence Applications to Constant False Alarm Rate (CFAR) Processing," *Record of the 1993 IEEE National Radar Conference*, 1993, pp. 275–280.

[24] Vannicola, V., R. Center, and A. Griffiss, "Expert System for Sensor Resource Allocation [radar application]," *Proceedings of the 33rd Midwest Symposium on Circuits and Systems*, 1990, pp. 1005–1008.

[25] Vannicola V. C., and J. A. Mineo, "Applications of Knowledge-Based Systems to Surveillance," *Proceedings of the 1988 IEEE National Radar Conference*, 1988, pp. 157–164.

[26] Wicks, M., W. Baldygo, Jr, and R. Brown, "Expert System Constant False Alarm Rate (CFAR) Processor," U.S. Patent 5,499,030, 1996.

[27] Capraro, C. T., et al., "Implementing Digital Terrain Data in Knowledge-Aided Space-Time Adaptive Processing," *Aerospace and Electronic Systems, IEEE Transactions on*, Vol. 42, 2006, pp. 1080–1099.

[28] Capraro, C. T., et al., "Improved STAP Performance Using Knowledge-Aided Secondary Data Selection," *Proceedings of the IEEE Radar Conference*, 2004, pp. 361–365.

[29] Capraro, G. T., et al., "Knowledge-Based Radar Signal and Data Processing: A Tutorial Review," *IEEE Signal Processing Magazine*, Vol. 23, 2006, pp. 18–29.

[30] Melvin, W., et al., "Knowledge-Based Space-Time Adaptive Processing for Airborne Early Warning Radar," *Aerospace and Electronic Systems Magazine*, Vol. 13, 1998, pp. 37–42.

[31] Wicks, M. C., et al., "Space-Time Adaptive Processing: A Knowledge-Based Perspective for Airborne Radar," *IEEE Signal Processing Magazine*, Vol. 23, 2006, pp. 51–65.

[32] Hall, D., and J. Linas, "An Introduction to Multisensor Data Fusion," *Proceedings of the IEEE*, Vol. 85, 1997, pp. 6–23.

[33] Nitzberg, R., *Radar Signal Processing and Adaptive Systems*, Norwood, MA: Artech House, 1999.

[34] Farina, A., *Antenna-Based Signal Processing for Radar Systems*, Norwood, MA: Artech House, 1992.

[35] Farina, A., and L. Timmoneri, "Real-time STAP Techniques," *Electronics & Communication Engineering Journal*, Vol. 11, 1999, pp.13–22.

[36] Brookner, E., Tracking and Kalman Filtering Made Easy, New York: John Wiley & Sons, 1998.

[37] Bergin, J. S., et al., "Evaluation of Knowledge-Aided STAP Using Experimental Data," *IEEE Aerospace Conference*, 2007, pp.1–13.

[38] Melvin, W. L., and G. A. Showman, "Performance Results for a Knowledge-Aided Clutter Mitigation Architecture," *Proceedings of the IET International Conference on Radar Systems*, Edinburgh, Scotland, 2007.

[39] Farin, A., A. De Maio, and S. Haykin, *The Impact of Cognition on Radar Technology*: Scitech Publishing, 2017.

[40] Zasada, D. M., J. J. Santapietro, and L. D. Tromp, "Implementation of a Cognitive Radar Perception/Action Cycle," *IEEE Radar Conference*, Cincinnati, OH, 2014, pp. 0544–0547.

[41] Guerci, J. R., R. M. Guerci, M. Ranagaswamy, J. S. Bergin, and M. C. Wicks, "CoFAR: Cognitive Fully Adaptive Radar," *IEEE Radar Conference*, Cincinnati, OH, 2014, pp. 0984–0989.

[42] Greenspan, M., "Potential Pitfalls of Cognitive Radars," *2014 IEEE Radar Conference*, 2014, pp. 1288–1290.

2
Optimum Multi-Input Multioutput Radar

2.1 Introduction

In this chapter, we derive the optimum multichannel input-output (MIMO) relations as a function of the radar channel (target, clutter, noise) characteristics. Note that *both* the transmit and receive functions are optimized—in contrast to conventional optimum/adaptive radar in which primarily the receiver is optimized (e.g., STAP). To streamline the mathematical nomenclature and presentation we adopt a matrix-vector algebraic formulation that is, of course, justified due to the finite bandwidth of all constituent signals and systems [1]. For a continuous-time development, see [1–4] for both the single-input, single-output (SISO) and MIMO cases, respectively.

We begin in Section 2.2 with the optimum transmit-receive configuration that maximizes output SINR, and thus the probability of detection for a prescribed false alarm rate for the additive Gaussian-colored noise AGCN case [5]. In Section 2.3, we address the clutter-dominant case, which yields a similar but distinct formalism for maximizing the signal-to-clutter ratio (SCR) as for the SINR case. Optimizing the transmit-receive configuration for optimal target identification is then addressed in Section 2.4. Finally, in Section 2.5, we discuss the constrained optimization problem.

As is typically the case when considering optimum versus adaptive signal processing, we will first assume clairvoyant knowledge of the underlying requisite channel characteristics. Later in Chapters 3 and 4, we relax this assumption and introduce adaptive and knowledge-aided methods, respectively, in an attempt to address the need to both simultaneously estimate channel characteristics as well as perform a joint MIMO transmit-receive optimization. However, the resulting performance is bounded by the optimum results derived in this chapter and thus can be used by a radar designer to ascertain the potential benefits of adopting adaptation on transmit.

Lastly, the examples selected in this chapter were chosen to illustrate the various instantiations of the fundamental optimum MIMO theory. However, they are by no means exhaustive, and do not fully represent the potential gains achievable—which are highly scenario dependent. Rather, it is hoped that the reader will be readily able to extend and adapt the methods presented herein to his or her application of interest.

2.2 Jointly Optimizing the Transmit and Receive Functions Case I: Maximizing SINR

Consider the basic radar block diagram in Figure 2.1. A generally complex valued, multidimensional transmitted signal **s** interacts with a target denoted by the target transfer matrix H_T. The resulting multidimensional echo is then received along with additive (and generally) colored Gaussian noise (AGCN) **n**. The vector-matrix formulation is completely general inasmuch as any combination of spatial and temporal dimensions can be represented.

2.2 Jointly Optimizing the Transmit and Receive Functions: Case I

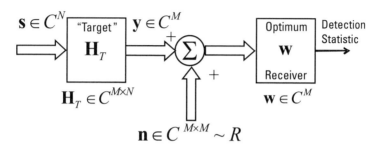

Figure 2.1 Fundamental multichannel radar block diagram for the AGCN case. Our objective is to design both the transmit (i.e., s) and receive (i.e., w) functions so as to maximize the output SINR given the channel characteristics.

For example, the N-dimensional input vector $\mathbf{s}$ could represent the N complex (i.e., in-phase and quadrature—I&Q [6]) samples of a single-channel transmit waveform $s(t)$; that is,

$$\mathbf{s} = \begin{bmatrix} s(\tau_1) \\ s(\tau_2) \\ \vdots \\ s(\tau_N) \end{bmatrix} \in C^N \qquad (2.1)$$

where C^N denotes the space of finite norm N-dimensional complex valued vectors.

The corresponding target transfer matrix $H_T \in C^{M \times N}$ (N transmit and M receive degrees-of-freedom (DoFs)) would thus contain the corresponding samples of the complex target impulse response $h_T(t)$, which for the causal linear time invariant (LTI) case would have the form [1]

$$H_T = \begin{bmatrix} h[0] & 0 & 0 & \cdots & 0 \\ h[1] & h[0] & 0 & \cdots & 0 \\ h[2] & h[1] & h[0] & \cdots & 0 \\ \vdots & & & \ddots & \vdots \\ h[N-1] & & & h[1] & h[0] \end{bmatrix} \qquad (2.2)$$

where, without loss of generality, we have assumed uniform time sampling; that is, $\tau_k = (k-1)T$, where T is a suitably chosen sampling interval [7]. Note also that without loss of generality, we have for both convenience and a significant reduction in mathematical nomenclature overhead, chosen $N = M$, the same number of transmit and receive DoFs (time, space, etc.). The reader is encouraged to, where desired, reinstate the inequality and confirm that the underlying equations derived throughout this chapter have the same basic form except for differing vector and matrix dimensionalities.

The formalism is readily extensible to the multiple transmitter, multiple receiver case (i.e., the conventional MIMO case). For example, if there are three independent transmit and receive channels (e.g., an active electronically scanned array or AESA), then the input vector **s** of Figure 2.1 would have the form

$$\mathbf{s} = \begin{bmatrix} \mathbf{s}_1 \\ \mathbf{s}_2 \\ \mathbf{s}_3 \end{bmatrix} \in C^{3N} \qquad (2.3)$$

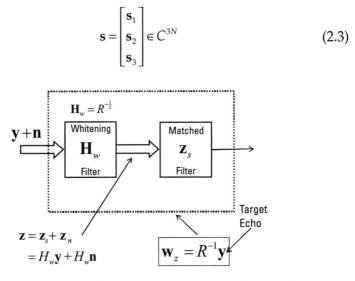

Figure 2.2 The optimum receiver for the AGCN case consists of a whitening filter followed by a white noise matched filter.

where $s_i \in C$ denotes the samples (as in (2.1)) of the transmitted waveform out of the ith transmit channel. The corresponding target transfer matrix would have the form

$$H_T = \begin{bmatrix} H_{11} & H_{12} & H_{13} \\ H_{21} & H_{22} & H_{23} \\ H_{31} & H_{32} & H_{33} \end{bmatrix} \in C^{3N \times 3N} \qquad (2.4)$$

where the submatrix $H_{i,j} \in C^{N \times N}$ is the transfer matrix between the ith receive and jth transmit channels for all time samples of the waveform.

These examples make clear that the matrix-vector input-output formalism is completely general and can accommodate whatever transmit-receive DoFs desired (e.g., fast-time (intrapulse), slow-time (interpulse), spatial, or polarimetric).

Returning to the task at hand; that is, optimizing the transmit and receive functions, we will find it convenient to work backwards from the receiver as follows: Irrespective of **s**, the receiver that maximizes output SINR for the AGCN case is the so-called whitening (or colored noise) matched filter, as shown in Figure 2.2 [5].

If $R \in C^{N \times N}$ denotes the total interference covariance matrix associated with **n**, which is further assumed to be Hermitian positive definite [8] (guaranteed in practice due to ever-present receiver noise [5]), the corresponding whitening filter is given by [5]

$$H_w = R^{-\frac{1}{2}} \qquad (2.5)$$

The whitening properties of H_w are readily verified as follows:

$$\begin{aligned} \mathrm{cov}(H_w \mathbf{n}) &= E\{(H_w \mathbf{n})(H_w \mathbf{n})'\} \\ &= E\{H_w (\mathbf{nn}') H_w'\} \\ &= H_w E\{\mathbf{nn}'\} H_w' \\ &= H_w R H_w' \\ &= R^{-\frac{1}{2}} R R^{-\frac{1}{2}} \\ &= I \end{aligned} \qquad (2.6)$$

where $E\{\ \}$ denotes the expectation operator [9].

The output of the linear whitening filter $\mathbf{z} \in C^N$ will consist of signal and noise components $\mathbf{z}_s$, $\mathbf{z}_n$, respectively, given by

$$\begin{aligned}\mathbf{z} &= \mathbf{z}_s + \mathbf{z}_n \\ &= H_w \mathbf{y} + H_w \mathbf{n} \\ &= H_w H_T \mathbf{s} + H_w \mathbf{n}\end{aligned} \quad (2.7)$$

where $\mathbf{y}_S \in C^N$ denotes the target echo as shown in Figure 2.1 (i.e., the output of H_T).

Since the noise has been whitened (identity covariance matrix), but is still Gaussian (normality is preserved under a linear—in this case full rank—transformation [5]), the final receiver stage consists of a white noise matched filter of the form

$$\mathbf{w}_z = \mathbf{z}_s \in C^N \quad (2.8)$$

The corresponding output SNR is thus given by

$$\begin{aligned}\text{SNR}_o &= \frac{|\mathbf{w}'_z \mathbf{n}_s|^2}{\text{var}(\mathbf{w}'_z \mathbf{z}_n)} \\ &= \frac{|\mathbf{z}'_s \mathbf{z}_s|^2}{\text{var}(\mathbf{z}'_s \mathbf{z}_n)} \\ &= \frac{|\mathbf{z}'_s \mathbf{z}_s|^2}{E\{\mathbf{z}'_s \mathbf{z}_n \mathbf{z}'_n \mathbf{z}_s\}} \\ &= \frac{|\mathbf{z}'_s \mathbf{z}_s|^2}{\mathbf{z}'_s E\{\mathbf{z}_n \mathbf{z}'_n\} \mathbf{z}_s} \\ &= \frac{|\mathbf{z}'_s \mathbf{z}_s|^2}{\mathbf{z}'_s \mathbf{z}_s} \\ &= |\mathbf{z}'_s \mathbf{z}_s|\end{aligned} \quad (2.9)$$

where, due to the whitening operation, $E\{\mathbf{z}_n \mathbf{z}'_n\} = I$.

In words, the output SNR is proportional to the energy in the whitened target echo. This fact is key to optimizing the input function: Choose **s** (the input) to maximize the energy in the whitened target echo,

$$\max_{\{s\}} |z'_s z_s| \qquad (2.10)$$

Substituting $z_s = H_w H_T s$ into (2.10) yields the objective function that explicitly depends on the input

$$\max_{\{s\}} |s'(H'H)s| \qquad (2.11)$$

where

$$H \triangleq H_w H_T \qquad (2.12)$$

Recognizing that (2.11) involves the magnitude of the inner product of two vectors **s** and $(H'H)s$, we readily have from the Cauchy-Schwarz theorem [10], the condition which **s** must satisfy to yield a maximum, namely s *must be collinear with* ,

$$(H'H)s_{opt} = \lambda_{max} s_{opt} \qquad (2.13)$$

That is the optimum input s_{opt} must be an eigenfunction of $(H'H)$ with associated maximum eigenvalue.

It is important to recognize that the above set of input-output design equations represent the absolute optimum that any combination of transmit-receive operations can achieve, and are thus of fundamental value to the radar systems engineer interested in ascertaining the value of advanced adaptive methods (e.g., adaptive waveforms, and transmit-receive beamforming).

Example 2.1 Multipath Interference

This example illustrates the optimum transmit receive configuration for maximizing output SINR in the presence of colored noise interference arising from a multipath broadband noise source. More specifically, for the single transmit-receive channel case, it derives the optimum transmit pulse modulation (i.e., pulse shape).

Figure 2.3 illustrates the situation at hand. A nominally broadband white noise source undergoes a series of multipath

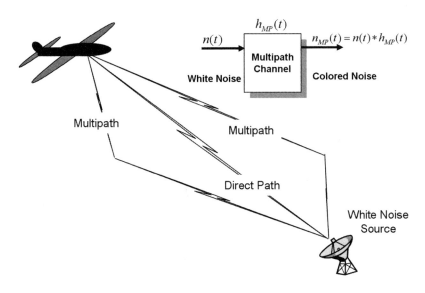

Figure 2.3 Illustration of colored noise interference resulting from a broadband (i.e., white noise) source undergoing multipath reflections.

scatterings that in turn colors the noise spectrum [11]. Assuming (for simplicity) that the multipath reflections are dominated by several discrete specular reflections, the resultant signal can be viewed as the output of a causal tapped delay line filter (i.e., an FIR filter [1]) of the form

$$h_{mp}[k] = \alpha_0 \delta[k] + \alpha_1 \delta[k-1] + \ldots + \alpha_{q-1} \delta[k-q-1] \quad (2.14)$$

that is driven by white noise. The corresponding input-output transfer $H_{mp} \in C^{N \times N}$ is thus given by

$$H_{mp} = \begin{bmatrix} h_{mp}[0] & 0 & \cdots & 0 \\ h_{mp}[1] & h_{mp}[0] & & \vdots \\ \vdots & & & 0 \\ h_{mp}[N-1] & \cdots & h_{mp}[1] & h_{mp}[0] \end{bmatrix} \quad (2.15)$$

In terms of the multipath transfer matrix, H_{mp}, the colored noise interference covariance matrix is given by

2.2 Jointly Optimizing the Transmit and Receive Functions: Case I

$$\begin{aligned}E\{\mathbf{nn'}\} &= E(H_{mp}\mathbf{vv'}H'_{mp})\\ &= H_{mp}E\{\mathbf{vv'}\}H_{mp}\\ &= H_{mp}H'_{mp}\\ &= R\end{aligned} \quad (2.16)$$

where the driving white noise source is a zero mean complex vector random variable with an identity covariance matrix,

$$E\{\mathbf{vv'}\} = I \quad (2.17)$$

Assuming a unity gain point target at the origin, $h_T[k] = \delta[k]$, yields a target transfer matrix $H_T \times C^{N \times N}$ given by

$$H_T = \begin{bmatrix} h_T[0] & 0 & \cdots & 0 \\ h_T[1] & h_T[0] & & \vdots \\ \vdots & & & 0 \\ h_T[N-1] & \cdots & h_T[1] & h_T[0] \end{bmatrix} \quad (2.18)$$
$$= I$$

While certainly a far more complex target model could be assumed, we wish to focus on the impact the colored noise has on shaping the optimum transmit pulse.

Figure 2.4(a, b) shows the in-band interference spectrum for the case when $\alpha_0 = 1$, $\alpha_2 = 0.9$, $\alpha_5 = 0.5$, $\alpha_{10} = 0.2$, all other coefficients are set to zero. The total number of fast time (range bin) samples was set to a short-pulse case of $N = 11$ (Figure 2.4(a)), and a long-pulse case of $N = 100$ (Figure 2.4(b)). Note that the multipath colors the otherwise flat noise spectrum. Also displayed is the spectrum of a conventional (and thus nonoptimized) LFM pulse with a time-bandwidth product ("$\beta\tau$") of 5 (Figure 2.4(a)) and 50 (Figure 2.4(b)), respectively [12, 13].

Given R from (2.16), the corresponding whitening filter H_w is given by

$$H_w = R^{-\frac{1}{2}} \quad (2.19)$$

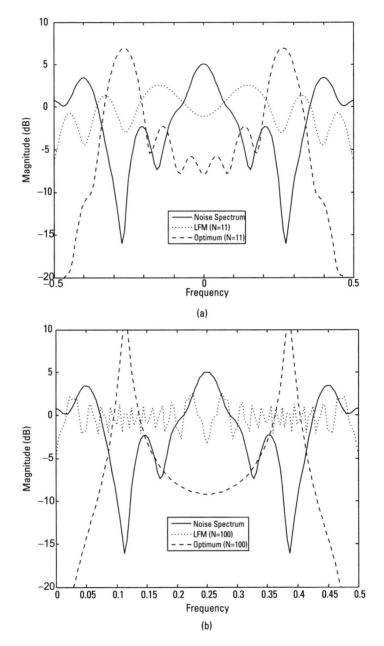

Figure 2.4 Spectra of the colored noise interference along with conventional and optimal pule modulations. (a) Short-pulse case where total duration for the LFM and optimum pulse are set to 11 range bins (fast-time taps), and (b) long-pulse case where total duration for the LFM and optimum pulse are set to 100 range bins. Note that in both cases the optimum pulse attempts to antimatch to the colored noise spectrum under the frequency resolution constraint set by the total pulse width.

2.2 Jointly Optimizing the Transmit and Receive Functions: Case I

Along with (2.18), the total composite channel transfer matrix H is thus given by

$$\begin{aligned} H &= H_w H_T \\ &= H_w \\ &= R^{-\frac{1}{2}} \end{aligned} \quad (2.20)$$

Substituting (2.20) into (2.13) yields

$$R^{-1}\mathbf{s}_{opt} = \lambda \mathbf{s}_{opt} \quad (2.21)$$

That is, the optimum transmit waveform is the maximum eigenfunction associated with the inverse of the interference covariance matrix. The reader should verify that this is also the minimum eigenfunction of the original covariance matrix R—and thus can be computed without matrix inversion.

Displayed in Figures 2.4(a, b) are the spectra of the optimum transmit pulses obtained by solving (2.21) for the maximum eigenfunction/eigenvalue pair for the aforementioned short- and long-pulse cases, respectively. Note how the optimum transmit spectrum naturally emphasizes those portions of the spectrum where the interference is weak—an intuitively satisfying result.

The SINR gain of the optimum short pulse, $SINR_{opt}$, relative to that of a nonoptimized chirp pulse, $SINR_{LFM}$, is

$$SINR_{gain} \triangleq \frac{SINR_{opt}}{SINR_{LFM}} = 7.0 \text{ dB} \quad (2.22)$$

While for the long-pulse case

$$SINR_{gain} \triangleq \frac{SINR_{opt}}{SINR_{LFM}} = 24.1 \text{ dB} \quad (2.23)$$

The increase in SINR for the long-pulse case is to be expected since it has finer spectral resolution and can, therefore, more precisely shape the transmit modulation to antimatch the interference.

Of course the unconstrained optimum pulse has certain practical deficiencies (such as poorer resolution and compression sidelobes) compared to a conventional pulse. We will revisit these issues in Section 2.5 where constrained optimization is introduced.

The above example is similar in spirit to the spectrum notching waveform design problem that arises when there are strong cochannel narrowband interferers present [14]. In this case it is not only desirable to filter out the interference on receive, but also to choose a transmit waveform that minimizes energy in the cochannel bands. The reader is encouraged to experiment with different notched spectra (i.e., different corresponding whitening filters) and applying (2.13) as was done in Example 2.1.

2.3 Jointly Optimizing the Transmit and Receive Functions Case II: Maximizing Signal-to-Clutter

Unlike the previous colored noise case in Section 2.2, clutter (i.e., channel reverberations) is a form of signal dependent noise [15, 16]—since the clutter returns depend on the transmit signal characteristics (e.g., transmit antenna pattern and strength, operating frequencies, bandwidths, and polarization).

In [3, 4], it was shown that the general clutter plus colored noise case does not allow for the same transmit-receive design decoupling that was exploited in Section 2.2. However, it is often the case that there is an effective decoupling between clutter and colored noise suppression. For example, in the presence of spatial noise jamming, spatial-only beamforming on receive is all that is required for jammer suppression [17, 18]—since the jamming is independent of the transmit configuration. This allows for the optimization of the transmit function to suppress clutter—which is the tact followed below.

Referring to Figure 2.5, the corresponding SCR at the input to the receiver is given by

$$\text{SCR} = \frac{\mathbf{y}'_T \mathbf{y}_T}{E\{\mathbf{y}'_c \mathbf{y}_c\}}$$

$$= \frac{\mathbf{s}'(H'_T H_T)\mathbf{s}}{\mathbf{s}' E\{H'_c H_c\} \mathbf{s}} \qquad (2.24)$$

2.3 Jointly Optimizing the Transmit and Receive Functions: Case II

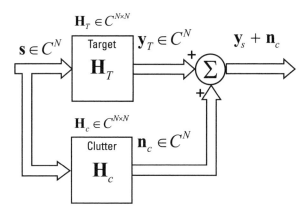

Figure 2.5 Radar signal block diagram for the clutter dominant case illustrating the direct dependency of the clutter signal on the transmitted signal.

where $H_c \in C^{N \times N}$ denotes the clutter transfer matrix, which is generally taken to be stochastic. Equation (2.24) is a generalized Rayleigh quotient [8] that is maximized when **s** is a solution to the generalized eigenvalue problem

$$(H_T' H_T)\mathbf{s} = \lambda (H_c' H_c)\mathbf{s} \qquad (2.25)$$

with corresponding maximum eigenvalue. When $H_c' H_c$ is positive definite, (2.25) can be converted to an ordinary eigenvalue problem of the form we have already encountered, specifically,

$$(H_c' H_c)^{-1}(H_T' H_T) = \mathbf{s}\lambda \qquad (2.26)$$

Before applying the above to the full-up ground clutter problem, we will first consider a simple, yet insightful example that illustrates the workings of to (2.25)–(2.26) as applied to the sidelobe target suppression problem.

Example 2.2 Sidelobe Target Suppression: Sidelobe Nulling on Transmit

Consider a narrowband $N = 16$ element ULA with half-wavelength interelement spacing and a quiescent pattern displayed in Figure 2.6. In addition to the desired target at a normalized angle of $\bar{\theta} = 0$, there are strong sidelobe targets at $\bar{\theta}_1 = -0.3$, $\bar{\theta}_2 = +0.1$,

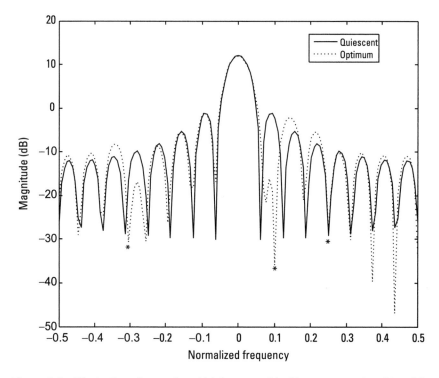

Figure 2.6 Illustration of proactive sidelobe target blanking on transmit achieved by maximizing the SCR. Note the presence of nulls in the directions of competing targets while preserving the desired main beam response.

$\bar{\theta}_3 = +0.25$ as shown Figure 2.6—where a normalized angle is defined as

$$\bar{\theta} \triangleq \frac{d}{\lambda}\sin\theta \qquad (2.27)$$

where d is the interelement spacing of the ULA and l is the operating wavelength (consistent units and narrowband operation assumed).

The presence of these targets could have been previously detected, and thus their directions known—especially if they were strong targets which are precisely the ones we are concerned about since their strong sidelobes could potentially mask weaker main lobe targets. With this knowledge, it is desired to minimize any energy from these targets leaking into the main beam detection of the target of interest by nulling on transmit—that is by placing transmit antenna pattern nulls in the directions of the unwanted targets.

For the case at hand, the (m, n)-th elements of the target and interferer transfer matrices are given, respectively, by

$$[H_T]_{m,n} = e^{j\varphi} \text{ (const.)} \tag{2.28}$$

$$[H_c]_{m,n} = \alpha_1 e^{j2\pi(m-n)\bar{\theta}_1} + \alpha_2 e^{j2\pi(m-n)\bar{\theta}_2} + \alpha_3 e^{j2\pi(m-n)\bar{\theta}_3} \tag{2.29}$$

where φ is an overall bulk delay (two-way propagation) that does not affect the solution to (2.25) and will thus be subsequently ignored; $[H_c]_{m,n}$ denotes the (m, n)-th element of the clutter transfer matrix and consists of the linear superposition of the three target returns resulting from transmitting a narrowband signal from the n-th transmit element and receiving it on the mth receive element of a ULA that utilizes the same array for transmit and receive [13, 17]. Note that in practice there would be a random relative phase between the signals in (2.29), which for convenience we have ignored but which can easily be accommodated by taking the expected value of the kernel $H'_c H_c$.

Solving (2.25) for the optimum eigenvector yields the transmit pattern that maximizes the SCR—which is the pattern also displayed in Figure 2.6. The competing target amplitudes were set to 40 dB relative to the desired target and 0 dB of diagonal loading was added to $H'_c H_c$ to improve numerical conditioning and allow for its inversion. Though this is somewhat arbitrary, it does provide a mechanism for controlling null depth—which in practice is limited by the amount of transmit channel mismatch [19]. Note the presence of transmit antenna pattern nulls in the directions of the competing targets as desired.

Example 2.3 Optimal Pulse Shape for Maximizing SCR

In this simple example, we rigorously verify an intuitively obvious result regarding pulse shape and detecting a point target in uniform clutter. Consider a unity point target, arbitrarily chosen to be at the origin. Its corresponding impulse response and transfer matrix are, respectively, given by

$$h_T[n] = \delta[n] \tag{2.30}$$

and

$$H_T = I_{N \times N} \tag{2.31}$$

where $I^{N,N}$ denotes the (N×N) identity matrix. For uniformly distributed clutter, the corresponding impulse response is of the form

$$h_c[n] \sum_{k=0}^{N-1} \tilde{\gamma}_k \delta[n-k] \tag{2.32}$$

where $\tilde{\gamma}_i$ denotes the complex reflectivity random variable of the clutter contained in the i-th range cell (i.e, fast-time tap). The corresponding transfer matrix is given by

$$\tilde{H}_c = \begin{bmatrix} \tilde{\gamma}_0 & 0 & 0 & \cdots & 0 \\ \tilde{\gamma}_0 & \tilde{\gamma}_0 & & & \\ \tilde{\gamma}_0 & \tilde{\gamma}_0 & \tilde{\gamma}_0 & & \\ \vdots & & & \ddots & \\ \tilde{\gamma}_{N-1} & \tilde{\gamma}_{N-2} & \tilde{\gamma}_{N-3} & \cdots & \tilde{\gamma}_0 \end{bmatrix} \tag{2.33}$$

Assuming that the $\tilde{\gamma}_i$ are i.i.d., we have

$$E\{\tilde{\gamma}_i^* \tilde{\gamma}_j\} = P_c \delta\{i-j\} \tag{2.34}$$

and thus

$$E\{[\tilde{H}_c' \tilde{H}_c]_{i,j}\} = \begin{cases} 0 & i \neq j \\ (N+1-i)P_c' & i = j \end{cases} \tag{2.35}$$

where $[\ ]_{i,j}$ denotes the (i, j)-th element of the transfer matrix. Note that (2.35) is also diagonal (and thus invertible), but with nonequal diagonal elements.

2.3 Jointly Optimizing the Transmit and Receive Functions: Case II

Finally, substituting (2.31) and (2.35) into (2.26) yields

$$E\{\tilde{H}'_c H_c\}^{-1} \mathbf{s} = \lambda \mathbf{s} \qquad (2.36)$$

where

$$E\{\tilde{H}'_c H_c\}^{-1} = \frac{1}{P_c}\begin{bmatrix} d_1 & 0 & \cdots & 0 \\ 0 & d_2 & & \\ & & \ddots & \\ 0 & & \cdots & d_N \end{bmatrix} \qquad (2.37)$$

and where

$$d_i \triangleq (N+i-1)^{-1} \qquad (2.38)$$

It is readily verified that the solution to (2.36) yielding the maximum eigenvalue is given by

$$\mathbf{s} = \begin{bmatrix} 1 \\ 0 \\ \vdots \\ 0 \end{bmatrix} \qquad (2.39)$$

That is, the optimum pulse shape for detecting a point target is itself an impulse. This should be immediately obvious since it is the shape that only excites the range bin with the target and zeros out all other range bin returns.

Of course, transmitting a short pulse is problematic in the real world (e.g., high peak power pulses) and thus an approximation to a short pulse in the form of a spread spectrum waveform (e.g., LFM) is often employed [12]. This example also makes clear that in the case of uniform random clutter, there is nothing to be gained by sophisticated pulse shaping for a point target, other than to maximize bandwidth (i.e., range resolution)—a well-known result rigorously proven by Manasse [20].

Example 2.4 Optimum Space-Time MIMO Processing for Clutter Suppression in Airborne MTI Radar

In this example, we tackle the MTI clutter suppression problem *proactively*—by judicious selection of our space-time (angle-Doppler) *transmit* patterns. This is in contrast to traditional space-time adaptive processing (STAP) methods that work only in the receiver [17]. We begin by briefly reviewing the airborne clutter problem for MTI radar. For a more detailed treatment, the reader is referred to [17, 21, 22].

Consider Figure 2.7, which depicts a sidelooking N-element ULA in motion. In the far field (long range/small depression angles) we see that the induced normalized Doppler shift on a differential patch of ground clutter at an angle relative to the array boresight is given by [17]

$$\overline{f}_d = \frac{2vT}{\lambda}\sin\theta \qquad (2.40)$$

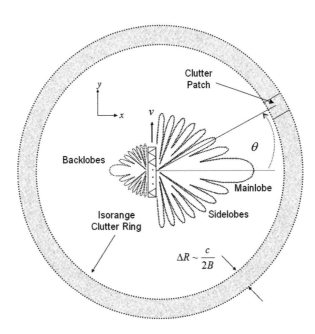

Figure 2.7 Illustration of clutter iso-range ring and angle-Doppler dependence for a side-looking ULA radar in constant velocity motion.

2.3 Jointly Optimizing the Transmit and Receive Functions: Case II

where v is the speed of the radar in units consistent with the operating wavelength λ and PRI T.

In addition to a spatially dependent Doppler shift, each clutter patch will have a spatially dependent reradiating intensity that depends on the transmit antenna pattern and the intrinsic reflectivity of the clutter [17, 21, 22]. In general, this pattern consists of both front lobe (main lobe and sidelobes) and backlobe radiation (see Figure 2.7). However, most airborne MTI radars are designed to insure that the back lobe radiation is significantly attenuated and can often be ignored [17]—which is the case considered here.

An illustration of the impact of uniform radar motion is provided in Figure 2.8. In the absence of platform motion ($v = 0$), the stationary clutter is concentrated along the zero Doppler contour for all normalized angles $\bar{\theta}$.

However, when $v \neq 0$ there is a linear relationship between normalized $\bar{f}_d$ Doppler and governed by (2.40) (see (2.43) below and [22]). As a result, the clutter energy is distributed along a line or clutter ridge as shown in Figure 2.8. Note that for the case illustrated, the antenna is aligned with direction of motion.

Assuming that the same ULA is used for both transmission and reception, and that a CPI is comprised of M pulses, the space-time narrowband clutter transfer matrix is of the form

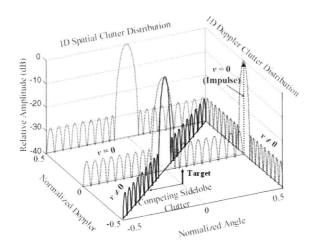

Figure 2.8 Illustration of the corresponding angle-Doppler clutter ridge arising from main lobe and sidelobe antenna patterns. Note that a target of interest generally must compete with either sidelobe or main lobe clutter leakage.

$$\tilde{H}_c = \begin{bmatrix} \tilde{H}_{C_{11}} & \tilde{H}_{C_{12}} & \cdots & \tilde{H}_{C_{1M}} \\ \tilde{H}_{C_{21}} & \tilde{H}_{C_{22}} & & \tilde{H}_{C_{2M}} \\ \vdots & & \ddots & \\ \tilde{H}_{C_{1M}} & \tilde{H}_{C_{M2}} & \cdots & \tilde{H}_{C_{MM}} \end{bmatrix} \in C^{NM \times NM} \qquad (2.41)$$

where the subblock $\tilde{H}_{C_{i,j}} \in C^{N \times N}$ is the array clutter transfer matrix between the ith and jth pulses. One at first might decide that the off-diagonal subblocks be set to zero since in ordinary pulsed radar operation the returns from different pulses at the same range bin do not overlap in time. However, we know that the receiver will be performing STAP in which the correlations between pulses at the same range bin will be computed and used to form the STAP filter [17]. Consequently, we will chose an input-output representation in which the crosscorrelations matter—and thus in general in which the channel matrix (2.41) generally contains nonzero off-block diagonal terms. Consequently, the (p, q) element of the (l, k) subblock is given by [23]

$$\left[H_{C_{lk}} \right]_{p,q} = \int_\Omega \tilde{\gamma}(\theta) e^{j2\pi[(p-q)+(l-k)\beta]\bar{\theta}} \qquad (2.42)$$

where $\tilde{\gamma}(\bar{\theta})$ is a complex random variable representing the intrinsic reflectivity corresponding to the infinitesimal clutter patch at a given (and fixed) range bin (assumed not to fluctuate during a CPI—Swerling 0 or 1 [6]), Ω denotes the angular field of view of the antenna (in this example it is taken to be ±90° relative to boresight), and β is the angle-Doppler coupling coefficient [17] that relates normalized Doppler to normalized angle,

$$\bar{f}_D = \beta \bar{\theta} \qquad (2.43)$$

where, from (2.40) and the half-wavelength spacing assumption, we have

$$\beta \triangleq \frac{d}{\text{PRI}} \qquad (2.44)$$

2.3 Jointly Optimizing the Transmit and Receive Functions: Case II

The corresponding random clutter kernel matrix $\tilde{H}'_c\tilde{H}_c$ is given by

$$\tilde{H}'_c\tilde{H}_c = \begin{bmatrix} \sum_r \tilde{H}'_{C_{r,1}}\tilde{H}_{C_{r,1}} & \sum_r \tilde{H}'_{C_{r,1}}\tilde{H}_{C_{r,2}} & \cdots & \sum_r \tilde{H}'_{C_{r,1}}\tilde{H}_{C_{r,M}} \\ \sum_r \tilde{H}'_{C_{r,2}}\tilde{H}_{C_{r,1}} & \sum_r \tilde{H}'_{C_{r,2}}\tilde{H}_{C_{r,2}} & & \\ \vdots & & \ddots & \\ \sum_r \tilde{H}'_{C_{r,M}}\tilde{H}_{C_{r,1}} & \cdots & & \sum_r \tilde{H}'_{C_{rM}}\tilde{H}_{C_{rM}} \end{bmatrix} \quad (2.45)$$

where the (p, q) element of the (l, k) subblock of the rth summation term is given by

$$\left[\tilde{H}'_{rl}\tilde{H}'_{rk}\right] = \iint_\Omega \langle \tilde{\gamma}^*(\bar{\theta}_1)\tilde{\gamma}(\bar{\theta}_2)\rangle e^{j2\pi\left[[(p-q)+(l-k)\beta]\bar{\theta}_2 - [(p-q)+(l-k)\beta]\bar{\theta}_1\right]} d\bar{\theta}_1 d\bar{\theta}_2 \quad (2.46)$$

For the special case of uncorrelated clutter, $E\{\tilde{\gamma}^*(\bar{\theta}_1)\tilde{\gamma}(\bar{\theta}_2)\} = G(\bar{\theta})\delta(\bar{\theta})\big|_{\bar{\theta} \triangleq \bar{\theta}_2 - \bar{\theta}_1}$, and thus (2.46) reduces to

$$\left[E\{H_{rl}H'_{rk}\}\right]_{p,q} = \int_\Omega G(\bar{\theta})e^{j2\pi\left[(p-q)+(l-k)\beta\right]\bar{\theta}} d\bar{\theta} \quad (2.47)$$

which is independent of r. For uniform clutter, $G(\bar{\theta}) = G = 1$ (choice of unity is arbitrary since the SCR can be set by adjusting the target strength below), and thus (2.47) simplifies to (to within an arbitrary phase constant)

$$\left[E\{H'_{rl}H'_{rk}\}\right]_{p,q} = \text{sinc}\left((p-q) + \beta(l-k)\right) \quad (2.48)$$

and thus

$$\left[\left[E\{H'_cH'_c\}\right]_{l,k}\right]_{p,q} = M\,\text{sinc}\left((p-q) + \beta(l-k)\right) \quad (2.49)$$

where $[[E\{H'_C H'_C\}]_{l,k}]_{p,q}$ denotes the (p, q)-th element of the (l, k)-th subblock of $E\{H'_C H_C\}$.

Assuming a deterministic point target at a normalized angle and Doppler $\bar{f}_{D_T}$, the corresponding target transfer matrix $H_T \in C^{NM \times NM}$ has the form

$$H_T = \begin{bmatrix} H_{T_{11}} & H_{T_{12}} & \cdots & H_{T_{1M}} \\ H_{T_{21}} & H_{T_{22}} & & H_{T_{2M}} \\ \vdots & & \ddots & \\ H_{T_{M1}} & H_{T_{M2}} & \cdots & H_{T_{MM}} \end{bmatrix} \quad (2.50)$$

where the (p, q)-th element of the (l, k) subblock is given by

$$\left[H_{T_{lk}} \right]_{p,q} = \kappa e^{j2\pi \left[(p-q)\bar{\theta}_T + (1-k)\bar{f}_T \right]} \quad (2.51)$$

Where κ is a positive constant used to set the SCR level.

We are now in a position to derive the optimal space-time MIMO transmit pattern that maximizes SCR by substituting (2.51) and (2.47) into (2.25). Doing so yields the space-time (angle-Doppler) transmit pattern shown in Figure 2.9, where we have set $\bar{\theta}_T = 0$, $\bar{f}_T = -0.25$, $N = M = 16$, and $\beta = 1$. Note that the clutter ridge has been prenulled, and a maximum response has been placed at the desired angle-Doppler cell corresponding to the target of interest.

While the above optimization is applicable to a single target engagement (i.e., one on one), there is often the need to simultaneously search for a number of targets. More specifically, we may wish to detect all targets present for a given look angle, say $\bar{\theta}_T$. In this case, the desired target transfer matrix would have elements of the form

$$\left[H_{T_{lk}} \right]_{p,q} = \sum_{i \in \Omega} \kappa e^{j2\pi \left[(p-q)\bar{\theta}_T + (1-k)\bar{f}_{T_i} \right]} \quad (2.52)$$

where, for example, the set of Doppler frequencies $\{\bar{f}_{T_i} : i \in \Omega\}$ might be chosen to span the entire unambiguous Doppler space (-0.5 to $+0.5$, sans the "0-Doppler" cell or cells corresponding to main

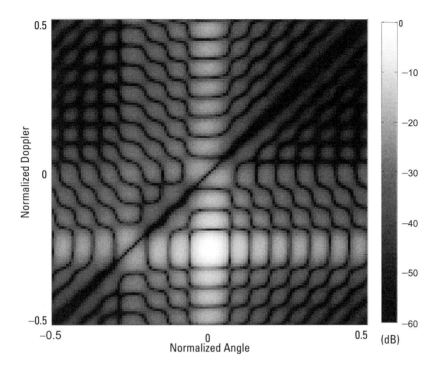

Figure 2.9 Resulting space-time (angle-Doppler) optimal transmit pattern for the uniform random clutter case arising from an airborne MTI radar scenario. Note that the two-dimensional transmit pattern can simultaneously place a peak on the desired target while notching out the clutter ridge.

beam clutter for the boresight-aligned side-looking radar case). Again, in practice the terms in (2.52) would contain random relative (but constant) phases that are easily accommodated by taking the expected value of $H'_T H_T$.

Figure 2.10 shows the resulting optimal angle-Doppler transmit pattern when there are multiple Doppler targets spanning the unambiguous nonmain-beam clutter region.

As with conventional STAP on receive, the above optimum space-time processing on transmit has many analogous practical difficulties—foremost of which are estimation of the channel transfer function and computational complexity. These and other issues are addressed in the subsequent chapters on *adaptive* and knowledge-aided MIMO radar. Also, the reader is referred to Corbell et al. [24] for an alternative synthesis of a space-time on trans-

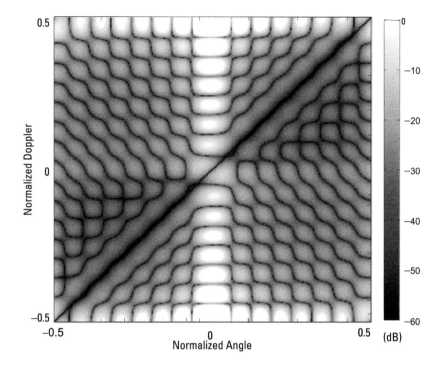

Figure 2.10 Angle-Doppler optimal transmit pattern that results for the multitarget (multi-Doppler) case.

mit transfer function (actually covariance matrix) that can be used to synthesize a space-time transmit pattern.

In the next section, we derive a mathematically similar set of transmit optimization equations for the target identification problem.

2.4 Optimum MIMO Target Identification

Consider the problem of determining target type when two known possibilities exist (the multitarget case is addressed later in this section). This can be cast as a classical binary hypothesis testing problem [5],

$$\begin{aligned}(\text{Target 1}) \quad & H_1 : \mathbf{y}_1 + \mathbf{n} = H_{T_1}\mathbf{s} + \mathbf{n} \\ (\text{Target 2}) \quad & H_2 : \mathbf{y}_2 + \mathbf{n} = H_{T_2}\mathbf{s} + \mathbf{n}\end{aligned} \quad (2.53)$$

2.4 Optimum MIMO Target Identification

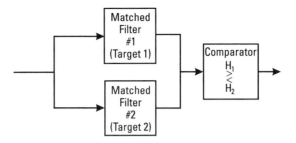

Figure 2.11 Optimal receiver structure for the binary (two target) hypothesis testing AGN problem.

where H_{T_1}, H_{T_2} denote the target transfer matrices for targets 1 and 2, respectively. For the AGN case, the well-known optimum receiver decision structure consists of a bank of matched filters, each tuned to a different target assumption, followed by comparator as shown in Figure 2.11 [5]. Note that the above presupposes that either Target 1 or 2 is present—but not both. Also, it has been tacitly assumed that a target present test has been conducted to ensure that a target is indeed present (i.e., binary detection test [5]). It should also be mentioned that the form of target ID considered herein is signal focused—and thus more utilitarian, versus the more esoteric and sophisticated pattern-recognition-based methods.

Figure 2.12 illustrates the situation at hand. If Target-1 is present, the observed signal $\mathbf{y}_1 + \mathbf{n}$ will tend to cluster about the #1 point in observation space—which could include any number of dimensions relevant to the target ID problem (e.g., fast-time, angle, Doppler, and polarization). The uncertainty sphere (generally ellipsoid for the AGCN case) surrounding #1 in Figure 2.12 represents the 1-sigma probability for the additive noise $\mathbf{n}$—similarly for #2. Clearly, if $\mathbf{y}_1$ and $\mathbf{y}_2$ are relatively well separated, the probability of correct classification is commensurately high.

Of significant note is the fact that $\mathbf{y}_1$ and $\mathbf{y}_2$ depend on the transmit signal $\mathbf{s}$, as shown in (2.53). Consequently, it should be possible to select an $\mathbf{s}$ that maximizes the separation between $\mathbf{y}_1$ and $\mathbf{y}_2$, thereby maximizing the probability of correct classification under modest assumptions regarding the conditional pdfs (e.g., unimodality),

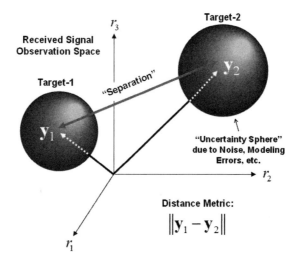

Figure 2.12 Illustration of the two-target ID problem. The goal of the joint transmitter-receiver design is to maximally separate the received signals in observation space, that in turn maximizes the probability of correct classification for the additive unimodal monotonic distributed noise case (e.g., AGN).

$$\max_{\{s\}} |d'd| \qquad (2.54)$$

where

$$\begin{aligned} d &\triangleq y_1 - y_2 \\ &= H_{T_1} s - H_{T_2} s \\ &= \left(H_{T_1} - H_{T_2} \right) s \end{aligned} \qquad (2.55)$$

where

$$H \triangleq H_{T_1} - H_{T_2} \qquad (2.56)$$

Substituting (2.55) into (2.54) yields

$$\max_{\{s\}} |s'H'Hs| \qquad (2.57)$$

which is precisely of the form (2.11), and thus has a solution yielding maximum separation given by

2.4 Optimum MIMO Target Identification

$$(H'H)\mathbf{s}_{opt} = \lambda_{max}\mathbf{s}_{opt} \qquad (2.58)$$

Equation (2.58) has an interesting interpretation: $\mathbf{s}_{opt}$ is that transmit input that maximally separates the target responses and is the maximum eigenfunction of the transfer kernel $H'H$ formed by the difference between the target transfer matrices (i.e., (2.56)).

Example 2.5 Two-Target Identification Example

Let $h_1[n]$ and $h_2[n]$ denote the impulse responses of targets #1 and #2, respectively, as shown in Figure 2.13. Figure 2.14 shows two different (normalized) transmit waveforms: (1) chirp; (2) optimum (per (2.58))—along with their corresponding normalized separation norms of 0.45 and 1, respectively (which corresponds to 6.9-dB improvement in separation). To determine the relative probabilities of correct classification for the different transmit waveforms, one would first need to set the SNR level (which fixes the conditional probability density functions (pdf) herein assumed to be cir-

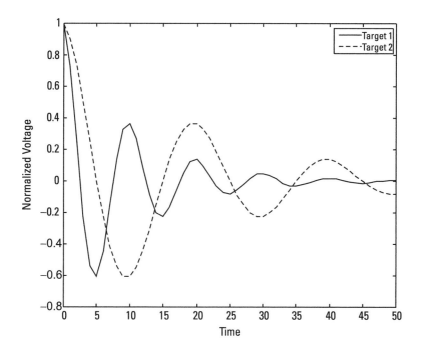

Figure 2.13 Target impulse responses utilized for the two-target identification problem.

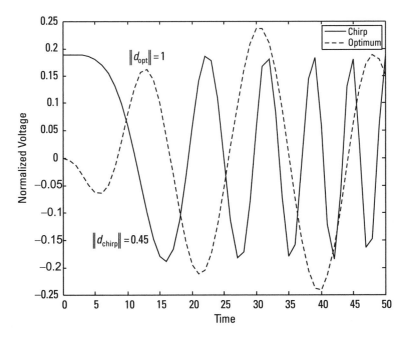

Figure 2.14 Transmit waveforms employed in the two-target identification example.

cular Gaussian), then measure the amount of overlap to calculate the probability [5].

An examination of Figure 2.15 reveals the mechanism by which enhanced separation is achieved. It shows the Fourier spectrum of $H(\omega) = H_{T_1}(\omega) - H_{T_2}(\omega)$, along with that of $S_{opt}(\omega)$. Note that $S_{opt}(\omega)$ places more energy in those spectral regions where $H(\omega)$ is large (i.e., regions where the difference between targets is large—again an intuitively appealing result).

While pulse modulation was used to illustrate the optimum transmit design equations, we could theoretically have used any transmit DOFs—polarization, Doppler, or all of the above. The choice clearly depends on the application at hand.

Multitarget Case

Figure 2.16 illustrates the multitarget case. To ensure that the L-target response spheres are maximally separated (an inverse sphere packing problem), we would like to jointly maximize the norms of the set of separations $\{\|\mathbf{d}_{ij}\| i = 1:L; j = i + 1: L\}$,

2.4 Optimum MIMO Target Identification

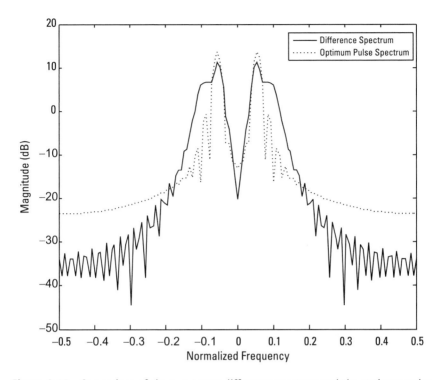

Figure 2.15 Comparison of the two-target difference spectrum and the optimum pulse spectrum. Note that the optimum pulse emphasizes those parts of the spectrum where the two targets differ the most.

$$\max_{\mathbf{s}} \sum_{i=1}^{L} \sum_{j=i+1}^{L} \left| \mathbf{d}'_{ij} \mathbf{d}_{ij} \right| \qquad (2.59)$$

Since, by definition, $\mathbf{d}_{ij}$ is given by

$$\begin{aligned} \mathbf{d}_{ij} &\triangleq \left(H_{T_i} - H_{T_l} \right) \mathbf{s} \\ &\triangleq H_{ij} \mathbf{s} \end{aligned} \qquad (2.60)$$

(2.59) can be rewritten as

$$\max_{\mathbf{s}} \mathbf{s} \left(\sum_{i=1}^{L} \sum_{j=i+1}^{L} H'_{ji} H_{ij} \right) \mathbf{s} \triangleq \mathbf{s}' K \mathbf{s} \qquad (2.61)$$

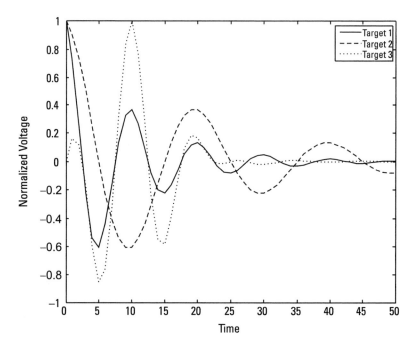

Figure 2.16 Target impulse responses utilized for the three-target identification problem.

Since $K \in C^{N \times N}$ is the sum of positive semidefinite matrices, it shares this same property and thus the optimum transmit input satisfies

$$K_{s_{opt}} = \lambda_{max} s_{opt} \qquad (2.62)$$

Example 2.6 Multitarget Identification Example

Figure 2.16 depicts the impulse responses of three different targets, two of which are the same as in Example 2.5. Solving (2.61)–(2.62) yields an optimally separating waveform whose average separation defined by (2.59) is 1.0, as compared to 0.47 for the chirp of Example 2.5—an improvement of 6.5 dB (slightly less than the previous example). As expected, the optimum waveform outperforms (in this case significantly) that of an unoptimized pulse such as the chirp.

The reader may have noticed the absence of a stochastic formulation of the target impulse responses. Clearly, in any real-world application, the target impulse responses are not precisely known a priori—and would thus have a stochastic formulation. Moreover, aspect angle uncertainties compound this lack of precision. This seeming oversight is easily justified as follows: (1) It is straightforward to replace the random kernel $\tilde{K}$ with its expected value $E\{\tilde{K}\}$, thereby maximizing the expected value of the average target separations; and (2) the more uncertainty in the target responses, the less value optimizing the input has since there is little to match to. Thus, a rule of thumb for when to optimize the transmit input for enhanced target ID is to limit its use to those situations with reasonably well-known target impulse responses (or equivalently transfer functions) of targets with good aspect angle knowledge (the case if under track), which further have significantly distinct characteristics relative to each other—an admittedly challenging set of criteria to meet in the real world, but not altogether impossible for certain scenarios where every dB counts.

2.5 Constrained Optimum MIMO Radar

Often in practice there are a number of practical considerations that may preclude transmitting the unconstrained optimum solutions developed so far. We will thus consider two cases of constrained optimization, namely Case 1: Linear Constraints; and Case 2: Nonlinear Constraints.

Case I: Linear Constraints

Consider the linearly constrained version of the input optimization problem:

$$\max_{\{\mathbf{s}\}} |\mathbf{s}'H'H\mathbf{s}| \quad (2.63)$$

$$\text{subject to: } G\mathbf{s} = 0 \quad (2.64)$$

where $G \in C^{Q \times N}$. To avoid the overly constrained case, it is assumed that $Q < N$. For example, the rows of G could represent steering vectors associated with known interferers such as unwanted targets and/or clutter discretes to which we wish to apply transmit nulls.

Equation (2.64) defines the feasible solution subspace for the constrained optimization problem. It is straightforward to verify that the projection operator

$$P = I - G'(GG)^{-1} G \qquad (2.65)$$

projects any $x \in C^N$ into the feasible subspace [25]. Thus, we can first apply the projection operator then perform an unconstrained subspace optimization to obtain the solution to (2.63) to (2.64),

$$\max_{\{s\}} |s'P'H'HPs| \qquad (2.66)$$

From which it is readily apparent that the constrained optimum transmit input satisfies

$$P'H'HPs_{opt} = \lambda_{max} s_{opt} \qquad (2.67)$$

Example 2.7 Prenulling on Transmit

If there are known directions for which it is desired not to transmit (unwanted targets, clutter discrete, keep-out zones, etc.), it is possible to formulate a linearly constrained optimization accordingly.

Assume that there is a desired target at $\bar{\theta}_T$, as well as two keep-out angles (normalized) $\bar{\theta}_{I_1}$ and $\bar{\theta}_{I_2}$. The corresponding elements of the target transfer matrix, assuming an N-element ULA, are thus given by

$$[H_T]_{m,n} = e^{j2\pi(m-n)\bar{\theta}_T} \qquad (2.68)$$

where $[H_T]_{m,n}$ denotes the (m, n)-th element of the target transfer matrix.

The keep-out constraints have the form

2.5 Constrained Optimum MIMO Radar

$$0 = \mathbf{Gs}$$
$$= \begin{bmatrix} \mathbf{s}'_{I_1} \\ \mathbf{s}'_{I_2} \end{bmatrix} \mathbf{s} \qquad (2.69)$$

where

$$\mathbf{s}'_{I_k} = \begin{bmatrix} 1 \\ e^{j2\pi \bar{\theta}_{I_k}} \\ \vdots \\ e^{j2\pi(N-1)\bar{\theta}_{I_k}} \end{bmatrix} \qquad (2.70)$$

Figure 2.17 shows the resulting constrained optimum transmit pattern for the case when $\bar{\theta}_T = 0$, $\theta_{I_1} = -0.25$, $\theta_{I_2} = 0.4$. As ex-

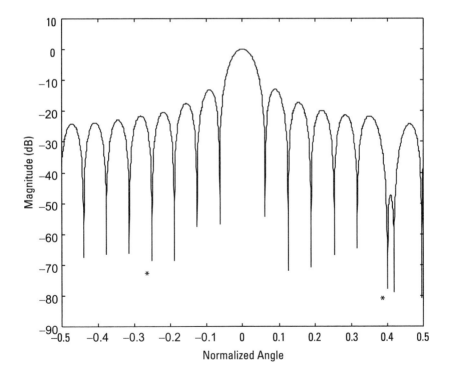

Figure 2.17 Example of a linearly constrained optimization in which two interferers are removed via the projection optimization approach.

pected a peak is placed in the desired target direction with nulls simultaneously placed in the keep-out directions.

Case II: Nonlinear Constraints

In practice there are other generally nonlinear constraints that may arise. One family of such constraints relates to the admissibility of transmit waveforms—such as the class of constant modulus and stepped frequency waveforms [12] to name a few.

For example, if it is desired to transmit a waveform that is nominally of the LFM type (or any other prescribed type), but which is allowed to modestly deviate to better match the channel characteristics, then the nonlinear constrained optimization has the form

$$\max_{\{s\}} |s'H'Hs| \qquad (2.71)$$

$$\text{subject to: } \|s - s_{LFM}\| \leq \delta \qquad (2.72)$$

The above and similar problems cannot generally be solved in closed form. However there are approximate methods that can yield satisfactory results—we will consider two such methods which are based on very different approaches. These simpler methods could form the basis of more complex methods—such as seeding nonlinear search methods.

Relaxed Projection Approach

Figure 2.18 depicts the constrained optimization problem of (2.71) to (2.72). It shows the general situation in which the unconstrained optimum solution does not reside within the constrained (i.e., admissible) subspace Ω. In this particular case, the admissible subspace is a convex set [26], defined as

$$\Omega = \{s : \|s - s_{LFM}\| \leq \delta\} \qquad (2.73)$$

From Figure 2.18 it is also immediately evident that the admissible waveform that is closest (in a normed sense) to the unconstrained optimum lies on the surface of Ω along the direction , which is the unit norm vector that points from s_{LFM} to s_{opt},

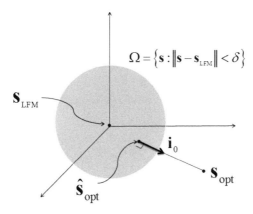

Figure 2.18 Illustration of a constrained optimization in which the signal should lie within a subspace (in this case convex) defined to be close to a prescribed transmit input (in this case an LFM waveform). The optimum relaxed projection is that point closest to the unconstrained optimum but which still resides in the subspace.

$$\mathbf{i}_o \triangleq \frac{\mathbf{s}_{opt} - \mathbf{s}_{LFM}}{\|\mathbf{s}_{opt} - \mathbf{s}_{LFM}\|} \quad (2.74)$$

Thus the constrained waveform that is closest in norm to $\mathbf{s}_{opt}$ is given by

$$\hat{\mathbf{s}}_{opt} = \mathbf{s}_{LFM} + \delta \mathbf{i}_o \quad (2.75)$$

Note that if δ is allowed to relax to the point where $\delta = \|\mathbf{s}_{opt} - \mathbf{s}_{LFM}\|$, then $\hat{\mathbf{s}}_{opt} = \mathbf{s}_{opt}$.

Example 2.8 Relaxed Projection Example

In this example, we impose an LFM similarity constraint on the multipath interference problem considered in Example 2.1. Specifically, in Figure 2.19, we plot the loss in SINR relative to the unconstrained long-pulse optimum solution originally obtained in Example 2.1 as a function of δ, which is varied between $0 \leq \delta \leq \|\mathbf{s}_{opt} - \mathbf{s}_{LFM}\|$. Note that for this example improvement generally monotonically increases with increasing δ, (except for a very small region near the origin), and that sizeable SINR improvements can be achieved for relatively modest values of the relaxation param-

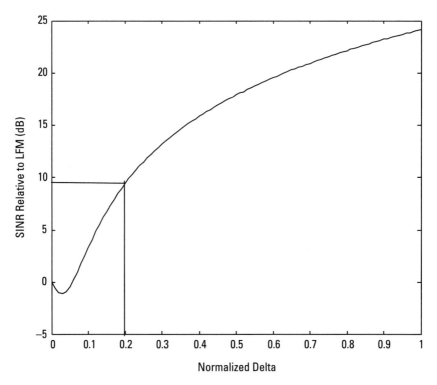

Figure 2.19 Illustration of the relaxed projection method for constrained optimization. The plot shows the SINR improvement relative to the unoptimized LFM waveform of Example 2.1 versus the normalized relaxation parameter d. Note that for even a modest relaxation of 20%, a nearly 10-dB gain in performance is achieved.

eter. In other words, a waveform with LFM-like properties can be constructed that still achieves significant SINR performance gains relative to an unoptimized LFM.

Figure 2.20 shows the spectra of the unoptimized LFM of Example 2.1, along with the unconstrained optimum and the relaxed projection pulse with a 20% relaxation parameter. Note how the relaxed pulse is significantly closer to the original LFM spectrum yet sill achieves nearly a 10-dB improvement in SINR relative to the LFM waveform.

Constant Modulus and the Method of Stationary Phase

As has become apparent from the previous examples, spectral shaping plays a key role in achieving matching gains. The method

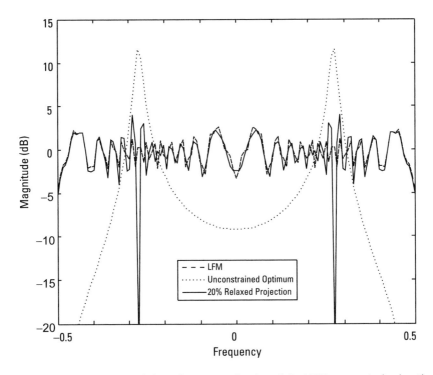

Figure 2.20 Comparison of the pulse spectra for the original LFM, unconstrained optimum, and 20% relaxed projection. Note how the relaxed pulse retains LFM-like spectral characteristics (and thus enhanced resolution for example) yet still achieves a 10-dB SINR improvement.

of stationary phase has been applied to the problem of creating a nonlinear frequency modulated (NLFM) pulse (and thus constant modulus) with a prescribed magnitude spectrum [1, 12].

Specifically, under fairly general conditions [1, 12], it is possible to relate instantaneous frequency ω of a NLFM waveform to time t, as follows: [1, 12]

$$\frac{1}{2\pi}\int_{-\infty}^{\omega}|S(\varpi)|^2\,d\varpi = k\int_0^t dt = kt \qquad (2.76)$$
$$t \in [0,T]$$

where $|S(\omega)|$ is the magnitude spectrum of the optimum pulse, and where we have assumed a constant modulus for the NLFM waveform resulting in a integral that is simply proportional to time (see

[1, 12] for the more general nonconstant modulus case)—as well as a finite and causal pulse.

Solving for ω was a function of t in (2.76) yields the frequency modulation, which will result in a transmit pulse with a magnitude spectrum equal to $|S(\omega)|$, to within numerical and other theoretical limitations [1, 12].

Example 2.9 Nonlinear FM (NLFM) to Achieve Constant Modulus

In this example, we use the method of stationary phase to design a constant modulus NLFM pulse that matches the magnitude spectrum of the optimum pulse derived for the multipath interference problem considered in Example 2.1.

Figure 2.21 shows the solution to (2.76); that is, ω versus t, along with the optimum pulse spectrum from Example 2.1 (long-pulse case). Note that as one would intuit, the frequency modulation dwells at those frequencies where peaks in the optimum pulse spectrum occur—and conversely note the regions in which the modulation speeds up to avoid frequencies where the optimum pulse spectrum has nulls or lower energy content.

The constant modulus NLFM waveform so constructed was able to achieve an output SINR that was within 6.0 dB of optimum, as compared with a 24-dB loss using an LFM waveform of same energy and duration.

It is natural to ask if a NLFM waveform with the same spectral magnitude as the optimum pulse (but not necessarily the same phase) will enjoy some (if not all) of the matching gains. For the steady-state case (infinite time duration) this is indeed true, since from Parseval's theorem [1] the output energy is related to only the spectral magnitudes (sans phase) of the input pulse and channel transfer function,

$$\frac{1}{2\pi}\int_{-\infty}^{\infty}|Y(\omega)|^2 d\varpi = \frac{1}{2\pi}\int_{-\infty}^{\infty}|H(\omega)|^2|S(\omega)|^2 d\omega \quad (2.77)$$

where $Y(\omega)$, $H(\omega)$, and $S(\omega)$ denote the Fourier transforms of the channel output, channel impulse response, and input pulse, re-

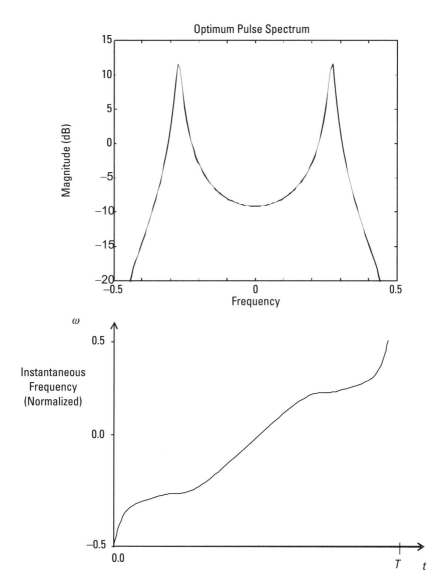

Figure 2.21 Illustration of the use of the method of stationary phase to create a constant modulus NLFM pulse whose spectral magnitude matches that of the optimum pulse. The NLFM pulse was able to achieve an output SINR that was within 6.0 dB of the optimum, as compared with a 24-dB loss using an LFM waveform of same energy and duration.

spectively. Note that the output energy in (2.77) depends on the spectral magnitude of the input pulse (steady state)—not the phase. Thus in theory, a NLFM waveform that exactly matches the optimum pulse magnitude spectrum will achieve the same matching gains in the steady state limit (infinite pulse duration).

Generalized Matched Subspace Projection (GMSP) Approach

The channel kernel matrix $K \triangleq E\{H'H\}$ has some important properties by virtue of its generally positive definite (or semidefinite) nature (see for example [8, 27]):

The eigenvalues of K generally satisfy

$$\lambda_{max} = \lambda_1 \geq \lambda_2 \geq \ldots \lambda_N \geq 0 \qquad (2.78)$$

The eigenfunctions of K, $\{\phi_i, i=1, \ldots, N\}$ form a complete orthonormal basis for C^N, i.e., if $\mathbf{x} \in C^N$, then

$$\mathbf{x} = \sum_{i=1}^{N} \alpha_i \varphi_i \qquad (2.79)$$

where

$$\alpha_i = \mathbf{x}' \varphi_i \qquad (2.80)$$

Next we define a matched subspace [28]:

Definition: A matched subspace Ω_{MS}, is defined as

$$\Omega_{MS} = \{\varphi_i : \lambda_i \geq \text{SINR}_{min}, \forall i\} \qquad (2.81)$$

In words, Ω_{MS} is the set of eigenfunctions of the channel kernel matrix with corresponding eigenvalues that are greater than (or equal to) a prescribed minimum SINR (or SCR) level.

Ω_{MS} thus defines the subspace in which a minimum SINR or SCR (i.e., SINR_{min}) is achieved. The purpose of such a subspace is to provide additional dimensions to the one-dimensional optimal solution so as to allow for the inclusion of other signal properties often imposed as constraints. Before illustrating the utility of such a subspace, we first define the matched subspace projection [28]:

Definition: A matched subspace projection (MSP) $\hat{s} \in C^N$ of a general signal $S \in C^N$ is defined as

$$\mathbf{s} \triangleq \sum_{[\varphi_i \in \Omega_{MS}]} \alpha_i \varphi_i \qquad (2.82)$$

where

$$\alpha_i = \mathbf{s}'\varphi_i \qquad (2.83)$$

The MSP has certain useful properties:

(P1): $\qquad \text{SINR}(\hat{\mathbf{s}}) \geq \text{SINR}_{min} \qquad (2.84)$

(P2): $\qquad \lim_{\|\varphi_i\| \to N} \hat{\mathbf{s}} = \mathbf{s} \qquad (2.85)$

where $\|\{\phi_i\}\|$ denotes the cardinality of the set of eigenfunctions. Proof of (P2) follows immediately from (2.79). (P1) can be proved as follows:

Proof of (P1)
From (2.82) we have

$$\hat{\mathbf{s}} \triangleq \sum_{i=1}^{L} \alpha_i \varphi_i \qquad (2.86)$$

Without loss of generality, we will further assume that the MSP signal is normalized,

$$\hat{\mathbf{s}}'\hat{\mathbf{s}} = 1 \qquad (2.87)$$

Substituting (2.86) into the expression for SINR (or SCR) yields

$$\hat{s}'K\hat{s} = \hat{s}K\left(\sum_{i=1}^{L} \alpha_i \varphi_i\right)$$

$$= \hat{s}' \sum_{i=1}^{L} \lambda_i \alpha_i \varphi_i$$

$$= \left(\sum_{i=1}^{L} \alpha_i^* \varphi_i\right)\left(\sum_{i=1}^{L} \lambda_i \alpha_i \varphi_i\right) \quad (2.88)$$

$$= \sum_{i=1}^{L} \lambda_i |\alpha_i|^2$$

Since (2.87) implies that

$$\sum_{i=1}^{L} |\alpha_i|^2 = 1 \quad (2.89)$$

we immediately have the desired result

$$\hat{s}'K\hat{s} = \sum_{i=1}^{L} \lambda_i |\alpha_i|^2 \geq \lambda_L \geq \text{SINR}_{\min} \quad (2.90)$$

We are now in a position to introduce a generalized matched subspace projection (GMSP) approach for constrained multichannel optimization [29]:

- Step 1: Determine whether optimizing the multichannel transmit input has any value. Examples chosen throughout this chapter have usually shown the value of optimizing the multichannel transmit input. In practice, this may not be the case.

  To simply check if the situation at hand can benefit from transmit optimization, it suffices to examine the eigenspectrum of the total channel kernel matrix K. For example, if the spectrum is flat (fully degenerate case), then any finite norm waveform in C^N will yield the same SINR. If, on the other hand, there is a significant eigenvalue spread then there is a commensurate potential for transmit optimization gain. Thus, some measure of the eigenvalue spread is indicated.

  One very common measure is the matrix condition number, which for positive definite matrices is the ratio of maximum to minimum eigenvalue. However care must be exercised in practice since only an estimate of K, namely , is available—which may have conditioning problems due to finite sample estimation effects [30]. In such circumstances, it may be necessary to perform a more sophisticated (and thus more computationally intensive) examination of the eigenspectrum.

A more practical approach is to simply test the nominal nonoptimum input(s) that would be transmitted if optimization was not to be performed. If $s_0 \in C^N$ is the nominal input, its SINR can be estimated from

$$\text{SINR}_0 \approx \frac{s_0' \hat{K} s_0}{s_0' s_0} \qquad (2.91)$$

which can be compared with an estimate of the optimum SINR (i.e., the max eigenvalue solution). If there is a significant difference, then proceed to Step 2.

- *Step 2:* Set the minimum SINR gain requirement; that is, SINR_{min}.

From Step 1, an estimate of the maximum SINR gain achievable is obtained. Since a constrained solution is likely to be less than this, one must establish a minimum SINR (i.e., SINR_{min}) below which it is presumably not worth the bother to optimize the multichannel transmit input. This value is used in the next step.

- *Step 3:* Calculate the matched subspace, that is, W_{MS}, from (2.81).

This will be the subspace into which the admissible waveforms will be projected.

- *Step 4:* Set the acceptable transmit criteria.

This step establishes the admissibility test for any candidate waveform. For example, in (2.73), the test measured the deviation of the proposed waveform from a standard LFM, however, any computable criteria could be used. Of course the more stringent the test, the less likely a transmit input will be found with significant matching gains—so care must be taken to make this criteria as flexible as possible!

- *Step 5:* Begin sequential search by first testing suitability of the max eigenvalue solution.

  First test to see if the max eigenvalue solution satisfies the constraints. If not, proceed to the next step.

- *Step 6:* Project nominal waveform(s) onto the matched subspace formed from the previous step plus the next dominant eigenvalue solution, then test for compliance.

  By adding an extra dimension to the matched subspace, we are guaranteed in general that the new projection is closer to the nominal waveform, yet will still have an SINR greater than $SINR_{min}$ from property P1. Repeat this step until the constraint is satisfied or no more eigenvalues exist satisfying $SINR_{min}$ —in which case the nominal (unoptimized) input would be transmitted.

Example 2.10   Matched Subspace Example

As with Example 2.8, we impose an LFM similarity constraint on the multipath interference problem considered in Example 2.1. In Figure 2.22(a, b), we plot both SINR and d (as defined in (2.72)) as a function of the cardinality of Ω_{MS} respectively. Note that for this example, the matched subspace approach is relatively slow to converge as compared to the relatively simple relaxation approach of Example 2.8. This, however, will not always be the case. In practice one might chose to run, in parallel, a number of constrained approaches and adaptively select the best performing algorithm. The interested reader is referred to [31] for a more detailed simulation involving site-specific clutter modeling.

2.6   Recent Advances in Constrained Optimum MIMO

Since the original publication of *Cognitive Radar: The Knowledge-Aided Fully Adaptive Approach,* there has been great interest in both cognitive radar in general and constrained optimal MIMO waveform design in particular. At the time of the writing of this second edition, there have been well over 100 peer-reviewed publications on cognitive radar since 2010! A significant fraction of these are focused on the constrained optimal waveform design problem.

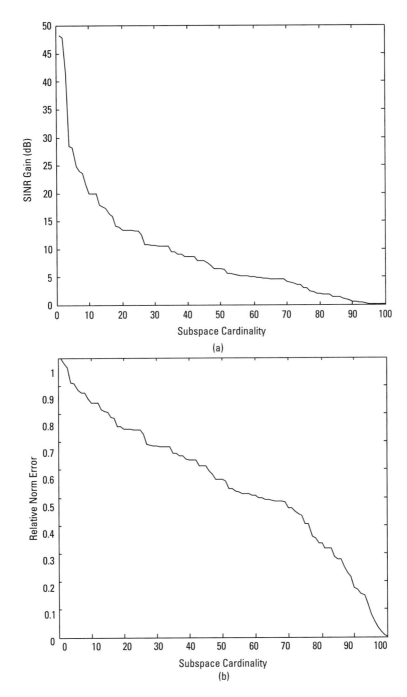

Figure 2.22 (a) SINR of projected solution relative to the unoptimized LFM waveform. (b) Relative norm error of the projection solution compared to the nominal LFM as a function of the dimension of the matched subspace.

While most entail a nonlinear optimization problem, and thus require numerical methods for their solution, a subset have been shown to lend themselves to convex optimization, which while still nonlinear, have much better-behaved numerical convergence properties [33–38]. That said, the techniques in this chapter should serve as a foundation for both the constrained and so-called unconstrained MIMO waveform optimization problem (although the latter still has finite energy and duration constraints).

References

[1] Papoulis, A., *Signal Analysis*, New York: McGraw-Hill, 1984.

[2] Grieve, P. G., and J. R. Guerci, "Optimum Matched Illumination-Reception Radar," U.S. Patent 5,175,552, 1992.

[3] Pillai, S. U., H. S. Oh, and J. R. Guerci, "Multichannel Matched Transmit-Receiver Design in Presence of Signal-Dependent Interference and Noise," *Proceedings of the 2000 IEEE Sensor Array and Multichannel Signal Processing Workshop*, 2000, pp. 385–389.

[4] Pillai, S. U., J. R. Guerci, and D. C. Youla, "Optimal Transmit-Receiver Design in the Presence of Signal-Dependent Interference and Channel Noise," *IEEE Transactions on Information Theory*, Vol. 46, 2000, pp. 577–584.

[5] van Trees, H. L., *Detection, Estimation and Modulation Theory: Part I*, New York: John Wiley & Sons, 1968.

[6] Barton, D. K., *Modern Radar System Analysis*, Norwood, MA: Artech House, 1988.

[7] Papoulis, A., *Circuits and Systems: A Modern Approach*, New York: Holt, Rinehart, and Winston, 1980.

[8] Horn, R. A., and C. R. Johnson, *Matrix Analysis*, Cambridge, United Kingdom: Cambridge University Press, 1990.

[9] Papoulis, A., and S. U. Pillai, *Probability, Random Variables, and Stochastic Processes*, New York: McGraw-Hill, 1991.

[10] Pierre, D. A., *Optimization Theory With Applications*, Courier Dover Publications, 1986.

[11] Guerci, J. R., and S. U. Pillai, "Theory and Application of Optimum Transmit-Receive Radar," *The Record of the IEEE 2000 International Radar Conference*, 2000, pp. 705–710.

[12] Cook, C. E., and M. Bernfeld, *Radar Signals*, New York: Academic Press, 1967.

[13] Richards, M. A., *Fundamentals of Radar Signal Processing*, New York: McGraw-Hill, 2005.

[14] Lindenfeld, M. J., "Sparse Frequency Transmit-and-Receive Waveform Design," *IEEE Transactions on Aerospace and Electronic Systems*, Vol. 40, 2004, pp. 851–861.

[15] van Trees, H. L., and E. Detection, *Modulation Theory: Part II*, New York: John Wiley & Sons, 1971.

[16] van Trees, H. L., *Detection, Estimation, and Modulation Theory: Radar-Sonar Signal Processing and Gaussian Signals in Noise*, Krieger Publishing Co., Inc., 1992.

[17] Guerci, J. R., *Space-Time Adaptive Processing for Radar*, Norwood, MA: Artech House, 2003.

[18] Rabideau, D. J., "Closed Loop Multistage Adaptive Beamforming," *Conference Record of the Thirty-Third Asilomar Conference on Signals, Systems and Computers*, 1999, pp. 98–102.

[19] Monzingo, R. A., and T. W. Miller, *Introduction to Adaptive Arrays*, SciTech Publishing, 2003.

[20] Manasse, R., "The Use of Pulse Coding to Discriminate Against Clutter," *Defense Technical Information Center (DTIC)*, vol. AD0260230, June 7, 1961.

[21] Klemm, R., *Principles of Space-Time Adaptive Processing*, IET, 2002.

[22] Ward, J., "Space-Time Adaptive Processing for Airborne Radar," *IEE Colloquium on Space-Time Adaptive Processing (Ref. No. 1998/241)*, 1998, p. 2.

[23] Guerci, J. R., et al., "Theory and Application of Optimum and Adaptive MIMO Radar," in *2008 IEEE Radar Conference*, Rome, Italy, 2008.

[24] Corbell, P. M., M. A. Temple, and T. D. Hale, "Forward-Looking Planar Array 3D-STAP Using Space Time Illumination Patterns (STIP)," *Fourth IEEE Workshop on Sensor Array and Multichannel Processing*, 2006, pp. 602–606.

[25] Gander, W., G. H. Golub, and U. von Matt, "A Constrained Eigenvalue Problem," *Linear Algebra Appl.*, Vol. 114, 1989, pp. 815–839.

[26] Youla, D. C., and H. Webb, "Image Restoration by the Method of Convex Projections: Part 1 Theory," *IEEE Transactions on Medical Imaging*, Vol. 1, 1982, pp. 81–94.

[27] Horn, R. A., and C. R. Johnson, *Topics in Matrix Analysis*, Cambridge University Press, 1991.

[28] Guerci, J. R., R. W. Schutz, and J. D. Hulsmann, "Constrained Optimum Matched Illumination-Reception Radar," U.S. Patent 5,146,229, 1992.

[29] Guerci, J. R., "Thinking Outside the Waveform: Input-Output Diversity," in *IEEE/IEE International Conference on Waveform Diversity and Design*, Kauai, HI, 2006.

[30] Golub, G. H., and C. F. Van Loan, *Matrix Computations*, 1996.

[31] Bergin, J. S., et al., "Radar Waveform Optimization for Colored Noise Mitigation," in *2005 IEEE International Radar Conference*, 2005, pp. 149–154.

[32] Farina, A., and F. A. Studer, "Detection with High Resolution Radar: Great Promise, Big Challenge," *Microwave Journal*, May 1991.

[33] Monga, V., "Constrained Optimization Paradigms for Adaptive, Cognitive Radar," The Pennsylvania State University, Air Force Research Laboratory, Technical Report, 2018.

[34] Kang, B., O. Aldayel, V. Monga, and M. Rangaswamy, "Spatio-Spectral Radar Beampattern Design for Co-existence with Wireless Communication Systems," IEEE Transactions on Aerospace and Electronic Systems, Vol. 55, No. 2, 2018, pp. 644–657.

[35] Jones, A. M., B. Rigling, and M. Rangaswamy, "Signal-to-Interference-Plus-Noise-Ratio Analysis for Constrained Radar Waveforms," IEEE Transactions on Aerospace and Electronic Systems, Vol. 52, 2016, pp. 2230–2241.

[36] Rangaswamy, M., A. Jones, and G. Smith, "Recent Trends and Findings in Cognitive Radar," in 2015 IEEE 6th International Workshop on Computational Advances in Multi-Sensor Adaptive Processing (CAMSAP), 2015, pp. 1–4.

[37] Setlur, P., and M. Rangaswamy, "Proximal Constrained Waveform Design Algorithms for Cognitive Radar STAP," in 2014 48th Asilomar Conference on Signals, Systems and Computers, 2014, pp. 143–147.

[38] Aubry, A., A. De Maio, M. Piezzo, M. M. Naghsh, M. Soltanalian, and P. Stoica, "Cognitive Radar Waveform Design for Spectral Coexistence in Signal-Dependent Interference," in 2014 IEEE Radar Conference, 2014, pp. 0474–0478.

Appendix 2.A: Infinite Duration (Steady State) Case

For the case when the observation interval becomes infinite (steady-state case), Fourier analysis can be employed to derive the optimal input. For example, for a SISO system, the energy in the whitened target echo is given by

$$\frac{1}{2\pi} \int_{-\infty}^{\infty} |H(\omega)|^2 |S(\omega)|^2 \, d\omega \qquad (2A.92)$$

Where $S(w)$ and $H(w)$ are the Fourier transforms of the input and composite channel transfer functions. Applying Schwarz's inequality to (2.92) yields the optimum choice of input, viz.,

$$|S(\omega)| \propto |H(\omega)| \qquad (2A.93)$$

Some of the earliest references to this result can be found in [32]. Note that while this result is strictly suboptimal for the finite duration case, it can be quite useful for: (1) A quick check on the potential gains of an channel-adaptive input; and (2) Providing an input which while suboptimal, is still nonetheless matched and often, in the case of waveform design, provides a better resolution solution.

3

Adaptive MIMO Radar

3.1 Introduction

Chapter 2 derived the optimal multidimensional transmit-receive (i.e., MIMO) design equations that assumed exact knowledge (deterministic and/or statistical) of the channel (target and interference). However, as those familiar with real-world radar are well aware, channel characterization in large part must be performed on-the-fly; that is to say, adaptively. This is simply a result of the dynamic nature of real-world targets and especially interference.

While a plethora of techniques have been developed for radar receiver adaptivity, estimating requisite channel characteristics for adapting the transmit function—especially for transmit-dependent interference such as clutter—is a relatively new endeavor.

In this chapter, we explore several approaches for addressing the adaptive MIMO optimization problem.

In Section 3.2, we introduce techniques for the case when the channel characteristics are independent of the transmit input—an example is additive colored noise jamming. Perhaps not surprisingly, given the transmit independence, the channel estimation techniques are essentially those often invoked in receive-only adaptivity (e.g., STAP [1]), although as will be shown the convergence properties can be quite different.

Section 3.3 introduces adaptive MIMO techniques for dynamic transmit array calibration—including the special case of cohere-on-target. This latter method enables the cohering of RF transmissions of distributed radars for a specific high-value target of interest.

In Section 3.4, we develop adaptive MIMO techniques for the case when the channel depends on the transmit selection. Of particular importance in this regard is clutter (i.e., channel reverberations) [2]. We will consider techniques that either use estimates of the clutter channel directly on transmit (e.g., STAP-on-transmit (STAP-Tx)) or indirectly through the use of orthogonal MIMO techniques [3–5]. This latter approach is applicable for systems that cannot practically implement true transmit adaptivity but nonetheless wish to achieve pseudo-transmit adaptivity by synthesizing an adaptive space-time transmit pattern in the receiver [6, 7].

Lastly, in Section 3.5, we will discuss recent advances in nonorthogonal MIMO probing techniques for channel estimation.

3.2 Transmit-Independent Channel Estimation

As mentioned previously, a multitude of techniques have been developed for the so-called transmit-independent case. A classic example is that of additive noise jamming [8]. For the case where no a priori knowledge is available, the baseline method of sample covariance estimation (and its many variants, such as diagonal loading, and principal components [1, 9, 10]) is often utilized. In addition to its statistical optimality properties (it is the maximum likelihood solution for the i.i.d. additive Gaussian noise case [11]), efficient parallel processing implementations have been developed facilitating its real-time operation [12].

Figure 3.1 depicts a common procedure for estimating additive, transmit-independent interference statistics. Specifically, the interference covariance matrix $R \in C^{N \times N}$ is approximated by $\hat{R} \in C^{N \times N}$, where

$$\hat{R} = \frac{1}{L} \sum_{q \in \Omega} \mathbf{x}_q \mathbf{x}_q' \qquad (3.1)$$

where $\mathbf{x}_q \in C^N$ denotes the N-dimensional receive array snapshot (spatial, spatiotemporal, etc.) corresponding to the qth independent temporal sample (e.g., a range or Doppler bin), and L denotes the number of i.i.d. samples selected from a suitable set of training samples to form the summation. As depicted in Figure 3.1, this training region is often chosen to be close in range to the range cell of interest (though there are many variants of this). If, moreover, the selected samples are Gaussian and i.i.d., then (3.1) can be shown to provide the maximum likelihood estimate of [11]. We illustrate this approach in the following example.

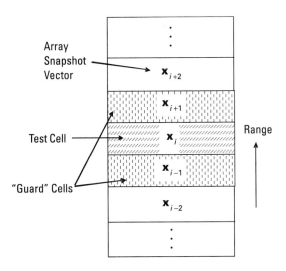

Figure 3.1 Illustration of a common method for estimating the interference statistics for the additive transmit-independent case.

Example 3.1 Adaptive Multipath Interference Mitigation

This is a repeat of Example 2.1 with the notable exception of unknown interference statistics that must be estimated on-the-fly. As a consequence, an estimate of the covariance matrix is used in (2.19) for the whitening filter rather than the actual covariance as was the case in Chapter 2.

Plotted in Figure 3.2 is the overall output SINR loss relative to the optimum for the short pulse case of Example 2.1 as a function of the number of independent samples used in (3.1). The results shown were based on 50 Monte Carlo trials (rms average) with a JNR of 50 dB and a small amount of diagonal loading to allow for inversion when the number of samples is less than 11 (positive semidefinite case).

It is interesting to note the rapid convergence and contrast this with SINR loss performance for adaptive beamforming—which is generally significantly slower. This is due to the fact that we are only estimating the single dominant eigenvalue/eigenvector pair. For an authoritative examination of principal components

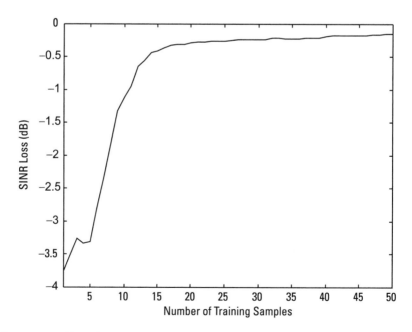

Figure 3.2 Effect of sample support on output SINR loss for the multipath interference scenario of Example 2.1.

estimation and convergence properties, the interested reader is referred to [13].

3.3 Dynamic MIMO Calibration

Perhaps the earliest instantiations of so-called MIMO radar techniques have their origins in transmit antenna array calibration [14, 15]. While techniques for estimating the receive array manifold using cooperative or noncooperative sources have been in existence for quite some time [14], methods for dynamically calibrating the transmit array manifold (such as active electronically scanned arrays (AESAs)) in situ are relatively recent developments [15].

An example of the utilization of MIMO techniques for dynamic calibration of an AESA radar is depicted in Figure 3.3. Orthogonal waveforms are simultaneously transmitted from each transmit-receive (T/R) site of an AESA (typically a single subarray AESA). A cooperative receiver then decodes each individual signal, calculates the relative phases (or time delays), then transmits this information back to the radar. By repeating this process for different orientations, a detailed look-up table for the transmit steering vectors can be constructed on-board the radar platform. This in situ approach is basically a necessity for very large AESAs in space since rigidity, which requires mass/weight, is not sufficient to maintain prelaunch calibration [15].

Example 3.2 MIMO Cohere-on-Target

An interesting special case of the above dynamic in situ calibration procedure is when transmit calibration is performed for a distributed radar focused on a single high-value target (HVT), as described by Coutts et al. [16].

Consider Figure 3.4, which depicts an airborne HVT that can be detected simultaneously by two geographically disparate radars. Given the HVT nature of the target, it is desired to have the two radars work coherently in order to maximize the overall SNR at each radar. To achieve on-target coherency, the two waveforms from each radar need to interfere constructively. To accomplish this,

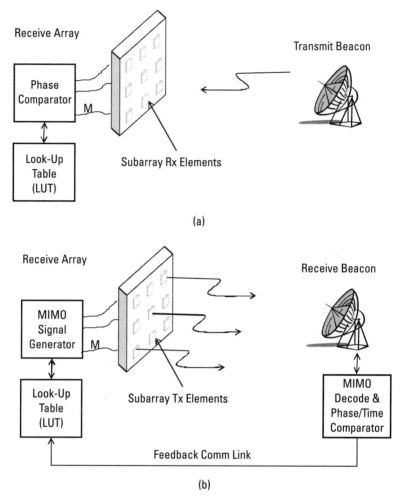

Figure 3.3 Illustration of a MIMO-based in situ calibration technique for an AESA radar [14, 15]: (a) conventional receive array calibration using a known in-band illuminator, and (b) the MIMO approach for calibrating the transmit array.

however, requires precise knowledge of the transmit pathways to a fraction of a wavelength [16]—essentially a dynamic calibration.

Drawing on MIMO-based calibration concepts, the requisite relative time delays between the two radars (as seen by the target) can be estimated by simultaneously transmitting orthogonal waveforms, which are then detected and processed in each radar as follows:

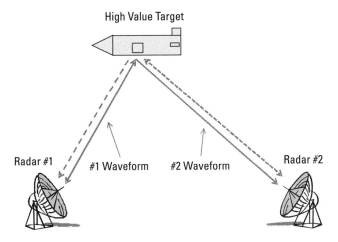

Figure 3.4 Illustration of the MIMO cohere-on-target approach for maximizing distributed radar performance [16].

- At each radar, the known one-way time delay to the target is subtracted from the total transit time for the sister radar (precise time synchronization is assumed). The remaining time delay is thus due to the first leg of the bistatic path (see Figure 3.4).
- By precompensating a joint waveform in each radar, the two waveforms can be made to cohere on the target—resulting in a 3-dB SNR boost (ideally). If the above procedure is repeated for N radars, as much as a $10 \log N$, dB gain in SNR is theoretically achievable.

While relatively straightforward to describe, the above procedure is replete with many real-world difficulties, including target motion compensation to a fraction of a wavelength and precise phase/timing stability.

3.4 Transmit-Dependent Channel Estimation

In the transmit-dependent case, the channel characteristics generally depend on the transmit spatiotemporal signals. Most notable in this regard is clutter. As was shown in Chapter 2, the space-time (angle-Doppler) clutter distribution is inextricably connected to the space-time transmit pattern of the radar. Indeed, in Chapter 2 we were able to use this dependence to our advantage and achieve clutter rejection via space-time transmit optimization (i.e., optimum MIMO) in contrast to a conventional radar that only adapts on receive (see Example 2.4). Of course, this presupposed knowledge of the clutter channel—something not generally available a priori.

In order to apply the STAP-on-transmit introduced in Chapter 2, a method for estimating on-the-fly the elements of the space-time clutter transfer matrix of equation (2.40) must be developed. Moreover, due to the dynamic nature of real-world radar scenarios (particularly from moving radars), a rapid method of estimation is required. For the two-dimensional (2-D) space-time example considered in Example 2.4, the individual transfer elements represent the correlation between individual transmit elements and pulses and receive elements. For example, the correlation between the second pulse transmitted from the third antenna element, and the fourth received pulse on the fifth receive antenna element.

If the same multielement array is used for transmit and receive, then it is possible to get an estimate of the requisite clutter channel by uniformly illuminating the clutter scene (spoiled transmit or transmission from one omni subarray/element), then processing the multichannel receive signals to produce an estimate of the space-time (angle-Doppler) spectrum. While not strictly optimum, treating this as a colored noise spectrum yields an optimum illumination spectrum per the design equations developed in Chapter 2 (to within transmit-receive array reciprocity—see below). The next example illustrates this STAP-on-transmit approach.

Example 3.2 STAP-Tx Example

Consider an airborne MTI radar employing an $N = 8$ element ULA in the usual sidelooking (no crabbing) configuration (see [1] for further details). Figure 3.5 shows the resulting angle-Doppler power spectrum corresponding to a uniform (omni) illumination over the usual ±90° FOV relative to boresight [1]. The example corresponds to the $\beta = 1$ case, with half-wavelength-spaced antenna elements and a coherent pulse interval (CPI) consisting of $M = 8$ PRIs [1]. Note the presence of the angle-Doppler clutter ridge in-

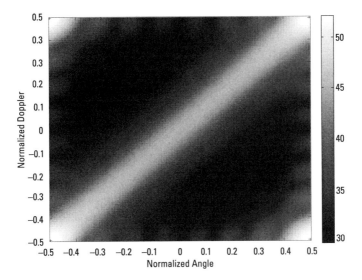

Figure 3.5 Resulting angle-Doppler power spectrum from a uniform (omni) transmit illumination. This power spectrum forms the basis for optimizing the space-time transmit pattern aimed at prenulling the clutter ridge.

duced by uniform ownship motion [1]. Ideally, we would like to have a space-time transmit pattern that prenulls this clutter ridge, thereby reducing the amount of residual competing clutter that must be canceled in the receiver.

Treating targets of interest as angle-Doppler point targets, the methodology developed in Chpater 2 for maximizing SINR in the presence of colored noise can be applied to obtain a transmit pattern that is antimatched to the clutter spectrum of Figure 3.5.

Let $\hat{R}_c \in C^{N \times N}$ denote the space-time covariance matrix estimated from the uniformly illuminated clutter. To alleviate finite sample estimation effects, it is assumed that a suitably chosen robust estimator was employed (such as diagonal loading and principal components). Assuming that the radar is in a scanning mode implies, for the present configuration, that targets of interest are in the mainbeam pointed along the boresight of the array. Additionally, we will assume that all unambiguous Doppler frequencies are of interest. Thus, the STAP-Tx approach would be of the form

$$\mathbf{W}_{Tx} = \hat{R}_c^{-1}\left(\mathbf{S}_1 + \ldots + \mathbf{S}_L\right) \in C^{NM} \qquad (3.2)$$

where $\mathbf{W}_T$ is the STAP-Tx vector of weights, and the steering vectors $\mathbf{S}_1, ..., \mathbf{S}_L$ are chosen to span the target Dopplers of interest along the boresight mainbeam (see Example 2.4).

Figure 3.6 shows the resulting space-time (angle-Doppler) transmit pattern corresponding to (3.2). Note the presence of a null along the clutter ridge and a passband along boresight across Doppler. Figure 3.5 compares the performance potential of this STAP-Tx approach with that of conventional STAP. Note the improvement in minimum detectable velocity (MDV)—a highly desirable feature, particularly for GMTI radar [1].

Adaptive beamforming on transmit can potentially introduce significantly more complexity and cost. Consequently, there has been sustained interest in approximating adaptive transmit performance by encoding the transmit DoFs in such a manner that they can be reconstructed in the receiver [3–5]. For better or worse, these techniques have become synonymous with MIMO radar—although as we have seen that MIMO radar has a much broader scope.

One such approach is the use of orthogonal waveforms that occupy the same frequency band in a manner precisely as that employed in code division multiple access (CDMA) [3–5]. A bank

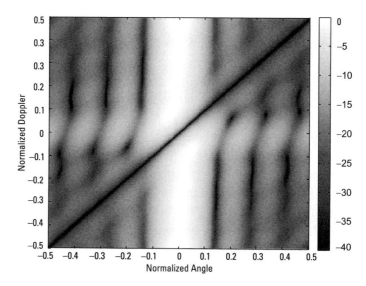

Figure 3.6 Adapted space-time (angle-Doppler) transmit pattern resulting from the STAP-on-transmit method.

of matched filters tuned to each of the transmit coded waveforms could then be used at each receive element to reconstruct each individual transmit signal—which in turn can then be used to estimate each matrix element of H_c—or form the basis for a higher-order STAP receiver (which is the most common MIMO radar architecture). An important consequence of the use of orthogonal transmit waveforms is the defocusing of the transmit antenna pattern (i.e., a spoiled beam). For the N transmit DoF case, the resultant loss in SNR is $10\log N$ (dB).

Of course, as with CDMA communications, the transmit DoFs cannot be perfectly reconstructed in the receiver since different transmit waveforms will essentially act as noise sources when match-filtering to one specific waveform. Consequently, the choice of waveform coding is still an active area of research [7]. For GMTI applications with extended range clutter (the norm), it has been shown that traditional noiselike CDMA waveforms have significant clutter leakage due their inherently poor integrated cross-correlation sidelobes [7].

A waveform coding scheme that avoids the above has been developed by Mecca et al. [6] and applied to GMTI by Bliss et al. [7]. It achieves orthogonality in the Doppler domain, and is thus referred to as Doppler division multiple access (DDMA). Assuming that the nominal clutter Doppler spectrum of interest is a small fraction of the unambiguous Doppler space, (which can be achieved with a sufficiently high pulse repetition frequency (PRF), it is possible to put a different slow-time linear phase code on each transmit aperture to modulate its corresponding clutter response to different portions of the unambiguous Doppler space [6, 7]. Relatively simple Doppler filtering can then be employed to separate the different clutter responses corresponding to each transmit DoF. We will explore in greater detail this approach in Example 3.2.

Example 3.2 DDMA MIMO STAP Clutter Mitigation Example for GMTI Radar

In this example we illustrate a particular instantiation of the aforementioned DDMA-based MIMO clutter mitigation example. It is a particularly efficient implementation (from a computational viewpoint) as encoding of the transmit DoFs is accomplished in

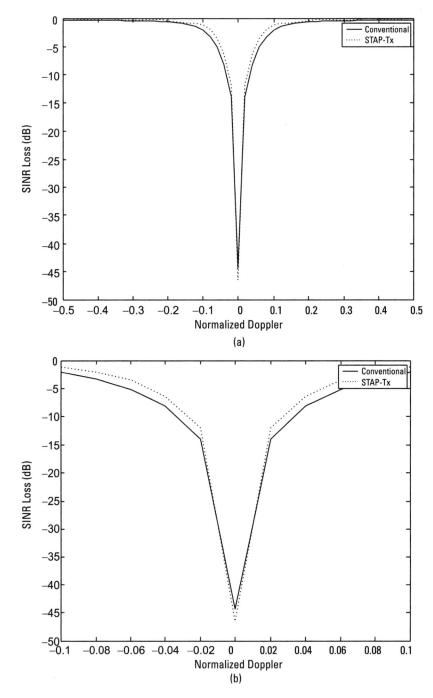

Figure 3.7 (a) Resulting SINR performance of the STAP-Tx method, and (b) blow-up of the low-Doppler regime showing the improved MDV performance.

the slow-time (Doppler) domain versus the more computationally demanding fast-time domain. Moreover, it exploits the significant rank-reduction properties achievable in post-Doppler adaptive processing methods often employed in STAP [1, 17].

Figure 3.8 provides a block diagram overview of the post-Doppler DDMA MIMO radar architecture selected for this example. One of the key tenets of STAP for airborne radar (which is the scenario considered here) is that ground clutter has an angle-Doppler dependency. In particular, this implies Doppler-dependent spatial nulling [1]. Referring to Figure 2.8 for example, we see that a target at a normalized Doppler of +0.25 would be competing with clutter arriving from a normalized angle of +0.25 (for the $\beta = 1$ case). This implies, to first order, that in a post-Doppler architecture one need only perform spatial-only nulling when forming the synthesized transmit antenna pattern in Step 4 above—as opposed to a more computationally intensive space-time (i.e., 2-D) transmit beamforming [1].

Again referring to Figure 3.8, we see that the proposed radar operating sequence consists of the following elements. For simplicity, we have assumed that the same three-element phased array is used for both transmit and receive.

1. Simultaneous transmission of coherent pulse trains from each transmit antenna elements/subarrays is performed. However, each transmit element has its own unique Doppler code (slow-time code). In this present example, we will assume a simple FDMA-like approach and apply a simple linear phase modulation to each element. Thus, each pulse interval will result in a differing phase progression across the transmit array that, in turn, will defocus the transmit antenna pattern. This spoiling of the transmit antenna pattern is a common characteristic of MIMO radar approaches utilizing orthogonal waveforms. This loss of transmit antenna gain directly impacts the SNR for a fixed-transmit ERP and coherent integration interval (we will account for this in the ensuing analysis).

2. At the receiver, the usual RF filtering, low-noise amplification, downconversion (IF), analog-to-digital conversion (ADC), and pulse compression is performed in each receive

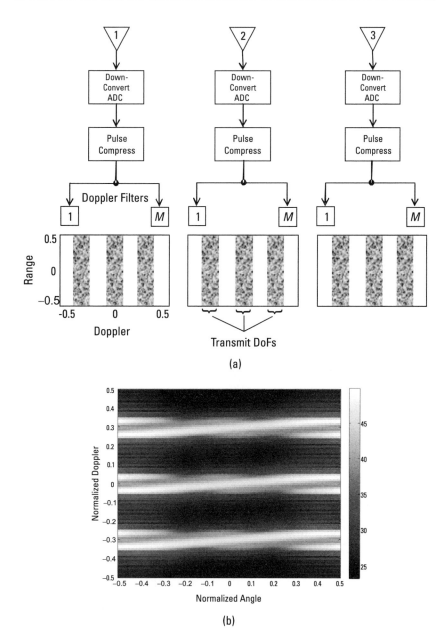

Figure 3.8 Illustration of the DDMA MIMO GMTI radar approach: (a) signal flow diagram illustrating the FDMA-like properties of the DDMA approach, and (b) the resulting angle-Doppler powers spectrum.

channel [18]. This produces a coherent set of M pulse returns (i.e., a coherent processing interval (CPI)), per range interval in each receive channel (in this case three channels).

3. Next an M-point DFT (or suitably chosen, possibly padded FFT) is performed to transform the CPI into the Doppler domain (in this case M Doppler filter bins), for each receive channel. The purpose of this stage is twofold: (1) decode the transmit DoFs by simple FDMA-like processing (see Figure 3.8), and (2) provide the inputs to a post-Doppler STAP algorithm (such as multibin post-Doppler STAP [1]).

4. At this stage the designer has a choice between implementing a full DoF MIMO-STAP filter or a factored approach.

3.5 Theoretical Performance Bounds of the DDMA MIMO STAP Approach

Let $\mathbf{X} \in C^{NM}$ denote the space-time receive-array snapshot complex vector RV corresponding to a range bin (fast-time sample) of interest [1]. For the $N = 3$ case, (we are assuming three transmit and receive elements using the same array), $\mathbf{X}$ is of the form

$$\mathbf{X} = \sum_{i=1}^{N_c} \tilde{\gamma}_{1,i} \mathbf{V}_{1,i} + \sum_{k=1}^{N_c} \tilde{\gamma}_{2,k} \mathbf{V}_{2,k} + \sum_{l=1}^{N_c} \tilde{\gamma}_{2,l} \mathbf{V}_{3,l} + \mathbf{n} \tag{3.3}$$

where we have employed a nominal statistically homogenous clutter model based on a Riemann sum approximation often employed in basic performance bound analyses [17]. Note that unlike the traditional coherent transmit antenna case, there are three distinct summation terms, each corresponding to the uniquely coded transmit DoFs—a distinguishing feature of the orthogonal MIMO approach. We will now discuss each of the constituent terms in (3.3) en route toward deriving an expression for the full DoF MIMO STAP covariance matrix used to gauge theoretical performance bounds.

N_c in (3.3) denotes the number of differential, uncorrelated clutter patches employed in the Riemann sum which, in this case, is taken to be ±90° relative to boresight of the sidelooking array (see [1] for further background and details regarding this choice and subsequent choices below). $\tilde{\gamma}_{q,i}$ denotes the complex reflectivity RV associated with the ith clutter patch and accounts for both intrinsic reflectivity and the energy reflected from the qth transmit antenna element. For the uncorrelated model employed in this analysis, we have

$$E\{\tilde{\gamma}_{q,i}\tilde{\gamma}^*_{q,k}\} = G_{q,i}\delta_{i,k} \tag{3.4}$$

where $\delta_{i,k}$ denotes the Kronecker delta function $\mathbf{V} \in [1,2,3]$, denotes the space-time steering vector corresponding to the qth transmit DoF and ith clutter patch, and is of the form

$$\mathbf{V}_{q,i} = \mathbf{b}_{q,i} \otimes \mathbf{a}_{q,i} \tag{3.5}$$

where $\otimes$ denotes the vector Kronecker product [19], and $\mathbf{b}_{q,i} \in C^M$ and $\mathbf{a}_{q,i} \in C^{N=3}$ denote the Doppler (temporal) and spatial steering vectors, respectively, for the qth transmit DoF and ith clutter patch and are of the form

$$\begin{aligned}\mathbf{a}_{q,i} &= e^{j2\pi\bar{\theta}_{q,j}}\mathbf{a}_i \\ &= e^{j2\pi\bar{\theta}_{q,j}}\begin{bmatrix} 1 \\ e^{j2\pi\bar{\theta}_i} \\ e^{j2\pi(2)\bar{\theta}_i} \end{bmatrix} \\ &= e^{j2\pi(q-1)\bar{\theta}_i}\begin{bmatrix} 1 \\ e^{j2\pi\bar{\theta}_i} \\ e^{j2\pi(2)\bar{\theta}_i} \end{bmatrix}\end{aligned} \tag{3.6}$$

where $\bar{\theta}_i$ is the normalized angle (2-D) to the ith clutter patch relative to the sidelooking array boresight (see [1] for further details) defined as

$$\bar{\theta}_i = \frac{d}{\lambda}\sin\theta_i \tag{3.7}$$

3.5 Theoretical Performance Bounds of the DDMA MIMO STAP Approach

where θ_i is the angle in radians to the ith clutter patch, d is the interelement spacing of the three-element uniform linear array (ULA), and λ is the operating wavelength (narrowband radar operation assumed [18]). Note that the prefix phase offset $e^{j2\pi\theta_{q,i}}$ in (3.6) accounts for the phase difference between transmit antennas, and that is the ordinary spatial receive array steering vector [1].

The MIMO STAP Doppler steering vector is defined as

$$\mathbf{b}_{q,i} = \mathbf{b}_q \odot \mathbf{b}_i \tag{3.8}$$

where $\odot$ denotes the Hadamard product (element-wise matrix/vector multiplication [19]), $\mathbf{b}_q \in \mathbb{C}^M$ is the intrinsic receive Doppler steering vector for the ith clutter patch, and $\mathbf{b}_q \in \mathbb{C}^M$ is the MIMO Doppler modulation used to code the qth transmit DoF [7]. These components are of the form

$$\mathbf{b}_i = \begin{bmatrix} 1 \\ e^{j2\pi\beta\bar{\theta}_i} \\ \vdots \\ e^{j2\pi(M-1)\beta\bar{\theta}_i} \end{bmatrix} \tag{3.9}$$

and

$$\mathbf{b}_i = \begin{bmatrix} 1 \\ e^{j2\pi\bar{f}_q} \\ \vdots \\ e^{j2\pi(M-1)\bar{f}_q} \end{bmatrix} \tag{3.10}$$

where we have adopted an FDMA-like Doppler frequency modulation approach suggested by [7] for the GMTI radar example. By suitable choice of normalized Doppler frequency offset f_q, and high-enough PRF, we can ensure that the modulated (shifted) clutter spectra corresponding to each transmit DoF is sufficiently separated and thus nonoverlapping. This allows for simple transmit DoF reconstruction in the receiver utilizing a post-Doppler approach described below.

The corresponding covariance matrix for the clutter plus noise signal defined by (3.3) is given by

$$E\{\mathbf{XX}'\} \triangleq R_c$$
$$= R_{c_{11}} + R_{c_{22}} + R_{c_{33}} + 2\text{Re}\{R_{c_{12}}\} + 2\text{Re}\{R_{c_{13}}\} + 2\text{Re}\{R_{c_{23}}\} \quad (3.11)$$

where

$$R_{c_{qq}} = \sum_{i=1}^{Nlc} G_{q,i} \mathbf{V}_{q,i} \mathbf{V}'_{q,i} \quad (3.12)$$

and

$$R_{c_{q,r}} = \sum_{i=1}^{Nlc} G_{q,r,i} \mathbf{V}_{q,i} \mathbf{V}'_{r,i} \quad (3.13)$$

where $G_{q,r,i} = E\{\tilde{\gamma}_{q,i}\tilde{\gamma}^*_{r,i}\}$ denotes the cross correlation between the qth and rth transmit signal echoes as seen by the ith clutter scatterer.

To calculate the ideal SINR for a target signal of interest (angle-Doppler) $\mathbf{S} \in C^{NM}$ achievable by a matched receiver, we can use the well-known relationship [1]

$$\text{SINR}_{opt} = \mathbf{S}' R_c^{-1} \mathbf{S} \quad (3.14)$$

where for the three-transmit MIMO DoFs case, is given by

$$\mathbf{S} = \mathbf{S}_1 + \mathbf{S}_2 + \mathbf{S}_3$$
$$= \mathbf{b}_{S_1} \otimes \mathbf{a}_{S_1} + \mathbf{b}_{S_2} \otimes \mathbf{a}_{S_2} + \mathbf{b}_{S_3} \otimes \mathbf{a}_{S_3} \quad (3.15)$$

where $\mathbf{S}_1, \mathbf{S}_2, \mathbf{S}_3$ denote the space-time steering vectors matched to the three transmit DoFs, and where

3.5 Theoretical Performance Bounds of the DDMA MIMO STAP Approach

$$\mathbf{b}_{S_q} = \mathbf{b}_q \odot \mathbf{b}_S \qquad (3.16)$$

where $\mathbf{b}_q \in C^M$ is the Doppler code for the qth transmit DoF and is given by (3.10), and $\mathbf{b}_S$ is the intrinsic Doppler steering vector corresponding to some target Doppler of interest. Similarly, $\mathbf{a}_{S_q} \in C^{N=3}$ is defined as

$$\mathbf{a}_{S_q} = e^{j2\pi(q-1)\bar{\theta}_S} \mathbf{a}_S \qquad (3.17)$$

where $\bar{\theta}_S$ is the angle to the target signal of interest and $\mathbf{a}_S$ is the intrinsic receive array angle steering vector corresponding to $\bar{\theta}_S$.

Figure 3.9 shows the full DoF optimum MIMO SINR versus Doppler for both the three-transmit DoF DDMA MIMO and the conventional coherent transmit antenna cases. A sidelooking (noncrabbing) long-range airborne radar case was assumed with an angle-Doppler coupling coefficient of $\beta = 0.08$ and a

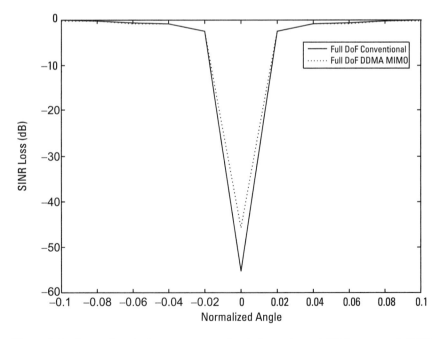

Figure 3.9 Comparison between a conventional three-element STAP array and MIMO DDMA. Note that even though the MIMO transmit array is incoherent, the clutter resolution is restored in the receiver. Note that this assumes that additional integration was performed to compensate for the loss of transmit antenna gain.

clutter-to-noise ratio (CNR) of 40 dB. Note that this choice of β corresponds to a rather high PRF for GMTI radar due to the need to ensure adequate separation between the modulated clutter returns for each transmit DoF. Each transmit antenna element/subarray is assumed to have an omni pattern from ±90° relative to boresight.

Note that while the DDMA MIMO approach does not cohere on transmit and thus has a broader clutter Doppler spread, it can recover the resolution in the receiver. This is a particularly attractive feature for applications for which cohering on transmit is difficult or essentially impossible, such as cooperative UAV illuminators [20]. This of course assumes that the loss in SNR due to the loss of transmit antenna gain is compensated for by either longer CPI's or greater time-bandwidth products.

Although this chapter introduced adaptivity-on-transmit, it drew on methods for adaptively estimating the channel that are grounded in well-established sample covariance methods. The next chapter discusses a more modern approach that exploits the proliferation of both environmental databases and knowledge-aided real-time computing.

3.6 Nonorthogonal MIMO Probing for Channel Estimation

Orthogonal MIMO probing techniques generally have the advantage of providing a rapid estimate of the signal-dependent channel (e.g., clutter). However, during the MIMO probe, primary radar functions such as search, track, and ID may be significantly degraded or altogether precluded. It is thus highly desirable to develop channel probing techniques that employ conventional space-time and generally nonorthogonal MIMO (NO-MIMO) waveforms.

In this section we introduce a new class of NO-MIMO probing techniques that can completely characterize the MIMO radar channel. We show that the inherent linearity of the Green's function representation of targets and clutter [21] allows for the use of nonorthogonal probing techniques. In particular, we show that a necessary and sufficient condition for complete MIMO channel characterization (in the absence of noise) is the use of N linearly independent transmit steering vectors, where N is the number of independent transmit DoFs.

Previously, research on MIMO radar focused on the advantages that can sometimes be afforded operating continuously in a MIMO mode [3, 5, 7, 22]. For example, in [23, 24], the additional virtual transmit DoFs were used to enhance clutter cancellation for GMTI and over-the-horizon (OTH) radar modes, respectively. In this section, we are focused on the usefulness of MIMO and NO-MIMO techniques as a means for probing the radar channel to obtain an estimate of the MIMO Green's function for clutter [25]. Armed with this knowledge, one can then utilize the capabilities of a fully adaptive radar (FAR) to optimize the transmit function characteristics; in other words, perform optimum MIMO (OptiMIMO) [26].

Precisely which MIMO probing technique to use in a particular circumstance is highly dependent on a multitude of factors, including the DoFs to be adaptively utilized on transmit and receive (fast-time, slow-time, spatial, etc.), the nature of the channel being estimated (e.g., clutter discretes versus distributed heterogenous clutter), and available radar resources. As discussed in [25, 26], the radar channel model for clutter and/or targets obeys a Green's function, which in a discretized representation is of the form

$$\mathbf{y} = H\mathbf{s} \tag{3.18}$$

where $\mathbf{s}$ is the complex *transmit* steering vector that is a concatenation of all relevant DoFs, H is the Green's function (generally stochastic) for the clutter or targets of interest, and is the received signal. Note that the dimensions of $\mathbf{s}$ and $\mathbf{y}$ need not be the same.

SISO Case

The SISO case refers to a radar with a single transmit and receive antenna. Time is thus the only DoF that could be optimized (fast and/or slow time). An estimate of the SISO channel Green's function fast-time response can be found by transmitting an impulse function (essentially the definition of a Green's function [21]). For any physically realizable radar, this is tantamount to transmitting a bandlimited impulse, which is of course equivalent to a sinc function. Fortunately, the LFM chirp closely approximates the sinc and can thus be used as a practically implementable alternative. Figure 3.10 shows a comparison of an ideal fast-time clutter Green's function response versus that obtained using an LFM using the RFView software for a site-specific application [27]. An illustration of the optimization of the fast-time waveform modulation for a given channel estimate can be found in [26, pp. 42–47].

Single-Input Multioutput (SIMO) Case

As with the SISO case, there is only one transmit DoF. So again, obtaining a channel estimate can be obtained by transmitting a bandlimited impulse waveform (or approximation). Optimizing the spatial or spatiotemporal DoFs in the multichannel receiver is a well-documented topic [28]. We will thus turn our attention to the MIMO case.

MIMO Case

In the spatial MIMO (or space-time) case, we seek to estimate the complete MIMO Green's function in order to jointly optimize the space (or space-time) transmit and receive functions [26, 28, 29]. Unlike the SISO or SIMO cases, if we wish to simultaneously estimate all the transmit-receive pathways, we must have a means of isolating the individual responses in the receiver. Needless to say, this is a potentially much more complicated procedure from both

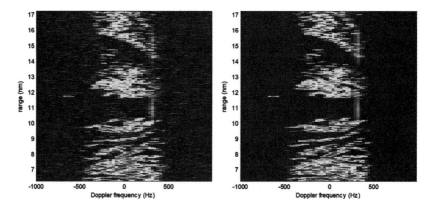

Figure 3.10 Comparison of (a) the ideal Green's function response, and (b) an estimate using an LFM waveform.

a radar hardware and processing point of view than conventional SIMO operations.

Table 3.1 lists some of the most common MIMO techniques that have been previously described in the literature along with their pros and cons. Just the salient features are listed. Each method comes with a host of other practical considerations including waveform diversity and VSWR (see [23] for a recent overview of these issues). (CDMA was one of the first MIMO techniques to be introduced [3, 30]. The key advantage is that all transmit-receive pathways can be probed simultaneously, thus minimizing the delay in characterizing the spatial MIMO response. This channel estimation approach has been shown to be quite effective in the look-up radar case in which there is minimal ground clutter response. However, for look-down airborne/spaceborne applications, the cochannel interference due to distributed clutter quickly becomes intolerable [23]. For radar channels that are slowly varying, time-division multiple access (TDMA) techniques might be adequate. Each transmit-receive pathway is probed sequentially in time. It should be noted that if only the spatial response is required and the ranges are relatively short, (i.e., short-range clutter), one can employ very short pulses with PRF, thereby minimizing radar timeline regret.

Again, which technique to employ is highly dependent on the operating scenario and constraint at hand. However, for some cases, certain techniques have gained popularity. For example, in GMTI applications for which distributed clutter is inherent,

Table 3.1
Pros and Cons of Orthogonal MIMO Techniques

MIMO Probe	Pros and Cons		
	Description	Pros	Cons
CDMA	Code-division multiple access	Simultaneous probing of all elements of H	Cochannel interference when distributed clutter present
TDMA	Time-division multiple access	Avoids cochannel interference problem	Requires slowly time-varying channel
FDMA	Frequency division multiple access	Simultaneous probing of all elements of H	Assumes channel is the same in different frequency subbands
DDMA	Doppler division multiple access	Simultaneous probing of all elements of H	Sacrifices usable Doppler space

DDMA techniques have been shown to be quite effective [23]. In the next section, we show that it is possible to fully estimate the channel without the use of orthogonal transmit waveforms.

Given the inherent linear nature of the Green's function, it is natural to ask if nonorthogonal waveforms could be employed to estimate the channel. The answer, as we will show, is affirmative. From (3.18) we have that the received response for a given transmit steering vector is

$$\mathbf{y}_1 = H\mathbf{s}_1 \qquad (3.19)$$

where H is the MIMO channel Green's function. In general, for the ith transmit steering vector,

$$\mathbf{y}_i = H\mathbf{s}_i \qquad (3.20)$$

For ease of presentation and without loss of generality, we will assume an equal number of transmit and receive DoFs, namely N. Thus, N transmit-receive pairs can be represented by the following set of simultaneous linear equations:

$$\mathbf{y} = S\mathbf{h}$$

$$\begin{bmatrix} \mathbf{y}_1 \\ \mathbf{y}_2 \\ \vdots \\ \mathbf{y}_N \end{bmatrix} = \begin{bmatrix} S_1 \\ S_2 \\ \vdots \\ S_N \end{bmatrix} \begin{bmatrix} \mathbf{h}_1 \\ \mathbf{h}_2 \\ \vdots \\ \mathbf{h}_N \end{bmatrix} \qquad (3.21)$$

where $\mathbf{h}_i$ is the column vector ($N \times 1$) of the ith row of H, and the ($N \times N^2$) matrix S_i is given by

$$S_i = \begin{bmatrix} \mathbf{s}_i^T & 0 & \cdots & 0 \\ 0 & \mathbf{s}_i^T & & 0 \\ \vdots & & \ddots & \vdots \\ 0 & 0 & \cdots & \mathbf{s}_i^T \end{bmatrix} \qquad (3.22)$$

where $\mathbf{s}_i^T$ $(1 \times N)$ denotes the transpose (without conjugation) of the ith steering vector. Note that $\mathbf{h}$, $(N^2 \times 1)$, which is the vector concatenation of the rows of H, represents the N^2 unknown elements of H. One can readily see that a necessary (not sufficient) condition is that at least N different transmit steering vectors must be sent. This ensures that S in (3.21) is at least $(N^2 \times N^2)$, and thus potentially invertible. Sufficiency is achieved if every row of S is linearly independent (not necessarily orthogonal). This leads to the following final set of necessary and sufficient conditions:

$$\text{rank}(S_i) = N, \quad \forall i = 1,\ldots,N$$
$$\|\mathbf{s}_i' \mathbf{s}_j\|^2 < \|\mathbf{s}_i\| \|\mathbf{s}_j\|, \quad \forall i,j : i \neq j \quad (3.23)$$
$$\|\mathbf{s}_i\| \neq 0, \quad \forall i = 1,\ldots,N$$

The nonzero norm requirement on S_i ensures that the rank of S_i is N, while Schwarz's inequality ensures that the steering vectors are not colinear (and thus linearly independent). Note that these conditions were developed without invoking any symmetry or reciprocity properties of H. In the absence of transmit-receive manifold and RF pathway errors, H can indeed be symmetric due to the electromagnetic reciprocity theorem [31]. In this case, the number of unknowns is almost cut in half.

The conditions of (3.23), while necessary and sufficient to ensure mathematical solubility, do not account for ever-present additive noise and interference. Thus, an extremely important area for future investigation is the selection of an optimal set of linearly independent steering vectors as a function of H and the channel interreference. For example, choosing the transmit steering vectors according to the maximum a posteriori criterion [11]

$$\max_{\{\mathbf{s}_1,\ldots,\mathbf{s}_N\}} f(H|\mathbf{y}_1,\ldots\mathbf{y}_N) \quad (3.24)$$

where $f(\ )$ denotes the posterior conditional probability density function of H conditioned on the observed measurements.

When more than N transmit steering vectors are used, (3.21) becomes an *overdetermined* set of linear equations. This provides additional opportunities for filtering in the presence of noise. For the additive Gaussian white nose case (AGWN), the least squares pseudo-inverse is optimum in a maximum likelihood sense; that is,

$$\hat{\mathbf{h}} = (S'S)^{-1} S'\mathbf{y} \qquad (3.25)$$

Using recursive least squares (RLS) methods, this estimate can be updated recursively and allows for fading memory to address the dynamically varying channel case [32].

Another interesting special case that satisfies the observability conditions is when $\mathbf{s}_i = \alpha_i \mathbf{e}_i$, where $\mathbf{e}_i$ is the Euclidean basis vector (α_i an arbitrary nonzero constant)

$$\mathbf{e}_i = \begin{bmatrix} 0 & \cdots & 0 & 1 & 0 & \cdots & 0 \end{bmatrix}^T \\ \uparrow i^{th} \text{ position} \qquad (3.26)$$

This corresponds to the aforementioned TDMA case in which only one transmit DoF is utilized at any given time. It is also a useful technique for calibrating the transmit array manifold for unknown errors [26]. Once the transmit array manifold is calibrated, other channel probing techniques could be employed that have less system regret.

In practice, physical insights into the problem at hand can significantly guide the selection of transmit steering vectors. For example, if we wish to characterize the spatial clutter channel that consists of a finite set of distributed large clutter discretes, it makes sense to choose a linear set of spatial transmit steering vectors that scan the relevant field-of-regard. This ensures that each clutter discrete has a good chance of receiving a reasonable amount of illumination energy, thus aiding the estimate of its Green's parameters in the presence of additive interference. This is the approach we will follow.

In [33], a new MIMO probing technique was introduced to assist in identifying large clutter discretes so that they could be prenulled via transmit adaptivity—either virtually via the transmit virtual array in the receiver or with the actual physical transmit antenna pattern. Prenulling large discretes, even if only partially effective, reduces the requirements on the receive-only adaptivity

3.6 Nonorthogonal MIMO Probing for Channel Estimation

for STAP [17, 28]. However, to achieve the requisite clutter discrete channel model, orthogonal transmit waveforms were utilized. Moreover, the technique in [33] did not explicitly estimate the composite Green's function H.

Guided by the discussion at the end of the previous section, we will choose a set of steering vectors that scan a field-of-regard for the purpose of estimating the MIMO Green's function response to a finite set of strong clutter discretes (strong point scatterers). This probing modality is also consistent with a wide area surveillance (WAS) mode that most surveillance radars naturally perform. In this way the regret of radar resources of MIMO probing of the channel is minimized or eliminated altogether.

The theoretical spatial Green's function response to a point target or clutter discrete, assuming a uniform linear array (ULA) with half-wavelength element spacing under a narrowband signal model, is given by [26]

$$[H_c]_{m,n} = \alpha_c e^{j2\pi(m-n)\bar{\theta}_c} \tag{3.27}$$

where $[H_c]_{m,n}$ denotes the (m,n)-th element of the spatial Green's function H_c and α_c, $\bar{\theta}_c$ denote the clutter discrete's amplitude and normalized angle-of-arrival (AoA), respectively [26].

When N_c spatially distinct point scatterers are present, the total MIMO Green's function is given by

$$[H_c]_{m,n} = \sum_{q=1}^{N_c} \alpha_{c_q} e^{j2\pi(m-n)\bar{\theta}_{c_q}} \tag{3.28}$$

In the following example, we will assume a finite set of clutter discretes are randomly distributed uniformly across the field-of-regard and are not known a priori.

Figure 3.11 shows both the distribution of random clutter discretes and the selection of nonorthogonal probing beams. Again, the selection of the beams was to ensure that each region that may contain a clutter discrete receives enough transmit energy to enable an effective estimate of the channel Green's function in the presence of additive noise/interference.

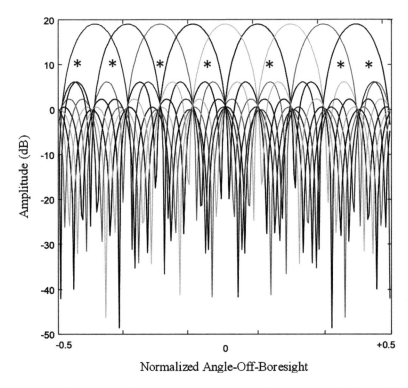

Figure 3.11 Illustration of the nine nonorthogonal probing beams used along with the angular locations of the clutter discretes (asterisks).

Figure 3.12 shows the L_2 matrix norm error between the estimated and actual channel Green's function versus CNR for a single discrete. We have assumed (for convenience) that each clutter scatterer is the same strength. As expected, performance improves with increasing CNR.

From [26], the optimum transmit steering vector satisfies the generalized eigenvector system given by

$$\lambda \left(E(H'_c H_c) + \sigma^2 I \right) \mathbf{s} = E(H'_T H_T) \mathbf{s}$$
$$\lambda \mathbf{s} = \left(E(H'_c H_c) + \sigma^2 I \right)^{-1} E(H'_T H_T) \mathbf{s} \qquad (3.29)$$

where $E(\ )$ denotes the usual expectation operator, and $\sigma^2 I$ denotes additive receiver noise. H_T is the desired target Green's function.

3.6 Nonorthogonal MIMO Probing for Channel Estimation 115

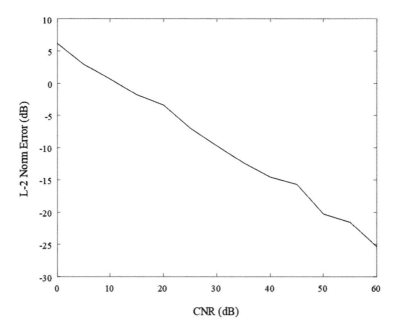

Figure 3.12 L-2 norm Green's function estimation error versus CNR.

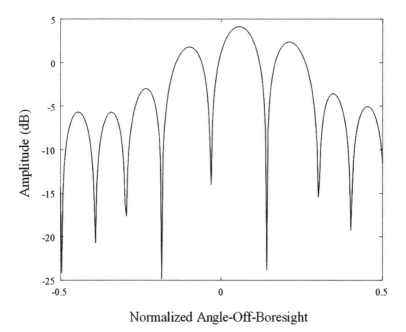

Figure 3.13 Adapted transmit pattern.

For the example at hand, we will assume it is a boresight aligned target.

Figure 3.13 shows that adapted optimum transmit pattern (7 nulls). Note that transmit nulls are placed in the direction of the discretes. Also note the squinting of the mainbeam off boresight due to the mainbeam nulling of one of the close-in discretes.

References

[1] Guerci, J. R., *Space-Time Adaptive Processing for Radar*, Norwood, MA: Artech House, 2003.

[2] Van Trees, H. L., *Detection, Estimation, and Modulation Theory: Radar-Sonar Signal Processing and Gaussian Signals in Noise*, Krieger Publishing Co., Inc., 1992.

[3] Bliss, D. W., and K. W. Forsythe, "Multiple-Input Multiple-Output (MIMO) Radar and Imaging: Degrees of Freedom and Resolution," presented at the *Conference Record of the Thirty-Seventh Asilomar Conference on Signals, Systems and Computers*, 2003.

[4] Forsythe, K. W., and D. W. Bliss, "Waveform Correlation and Optimization Issues for MIMO Radar," in *Conference Record of the Thirty-Seventh Asilomar Conference on Signals, Systems and Computers*, 2005, pp. 1306-1310.

[5] Forsythe, K. W., D. W. Bliss, and G. S. Fawcett, "Multiple-Input Multiple-Output (MIMO) Radar: Performance Issues," in *Conference Record of the Thirty-Seventh Asilomar Conference on Signals, Systems and Computers*, 2004, Vol. 1, pp. 310–315.

[6] Mecca, V. F., J. L. Krolik, and F. C. Robey, "Beamspace Slow-Time MIMO Radar for Multipath Clutter Mitigation," in *ICASSP 2008, IEEE International Conference on Acoustics, Speech and Signal Processing*, 2008, pp. 2313–2316.

[7] Bliss, D. W., K. W. Forsythe, S. K. Davis, et al., "GMTI MIMO Radar," presented at the 2009 International Waveform Diversity and Design Conference, IEEE, 2009.

[8] Monzingo, R. A., and T. W. Miller, *Introduction to Adaptive Arrays*, SciTech Publishing, 2003.

[9] Carlson, B. D., "Covariance Matrix Estimation Errors and Diagonal Loading in Adaptive Arrays," *IEEE Transactions on Aerospace and Electronic Systems*, Vol. 24, 1988, pp. 397–401.

[10] Haimovich, A. M., and M. Berin, "Eigenanalysis-Based Space-Time Adaptive Radar: Performance Analysis," *IEEE Transactions on Aerospace and Electronic Systems*, Vol. 33, 1997, pp. 1170–1179.

[11] Van Trees, H. L., *Detection, Estimation and Modulation Theory. Part I*, New York: Wiley, 1968.

[12] Farina, A., and L. Timmoneri, "Real-Time STAP Techniques," *Electronics & Communication Engineering Journal*, Vol. 11, 1999, pp. 13–22.

[13] Smith, S. T., "Covariance, Subspace, and Intrinsic Cramer-Rao Bounds," *IEEE Transactions on Signal Processing*, Vol. 53, 2005, pp. 1610–1630.

[14] Steinberg, B. D., and E. Yadin, "Self-Cohering an Airborne e Radio Camera," *IEEE Transactions on Aerospace and Electronic Systems*, Vol. AES-19, 1983, pp. 483–490.

[15] Guerci, J., and E. Jaska, "ISAT - Innovative Space-Based-Radar Antenna Technology," in *IEEE International Symposium on Phased Array Systems and Technology*, 2003, pp. 45–51.

[16] Coutts, S., K. Cuomo, J. McHarg, F. Robey, and D. Weikle, "Distributed Coherent Aperture Measurements for Next Generation BMD Radar," in *Fourth IEEE Workshop on Sensor Array and Multichannel Processing*, 2006, pp. 390–393.

[17] Ward, J., "Space-Time Adaptive Processing for Airborne Radar," *IEE Colloquium on Space-Time Adaptive Processing (Ref. No. 1998/241)*, 1998, p. 2.

[18] Richards, M. A., *Fundamentals of Radar Signal Processing*, McGraw-Hill, 2005.

[19] Horn, R. A., and C. R. Johnson, *Topics in Matrix Analysis*, Cambridge University Press, 1991.

[20] Guerci, J. R., M. C. Wicks, J. S. Bergin, P. M. Techau, and S. U. Pillai, "Theory and Application of Optimum and Adaptive MIMO Radar," presented at the 2008 IEEE Radar Conference, Rome, Italy, 2008.

[21] Greenberg, M. D., *Applications of Green's Functions in Science and Engineering*, Courier Dover Publications, 2015.

[22] Bliss, D. W., "Coherent MIMO Radar," presented at the 2010 International Waveform Diversity and Design Conference (WDD), 2010.

[23] Bergin, J. S., and J. R. Guerci, *Introduction to MIMO Radar*, Norwood, MA: Artech House, 2018.

[24] Mecca, V., J. Krolik, and F. Robey, "Beamspace Slow-Time MIMO Radar for Multipath Clutter Mitigation," in *IEEE International Conference on Acoustics, Speech and Signal Processing, ICASSP*, 2008, pp. 2313–2316.

[25] Guerci, J. R., J. S. Bergin, R. J. Guerci, M. Khanin, and M. Rangaswamy, "A New MIMO Clutter Model for Cognitive Radar," in *2016 IEEE Radar Conference (RadarConf)*, 2016, pp. 1–6.

[26] Guerci, J. R., *Cognitive Radar: The Knowledge-Aided Fully Adaptive Approach*, Norwood, MA: Artech House, 2010.

[27] *RFView(TM)*. Available: http://rfview.islinc.com.

[28] Guerci, J. R., *Space-Time Adaptive Processing for Radar*, Second Edition, Norwood, MA: Artech House, 2014.

[29] Guerci, J. R., R. M. Guerci, M. Ranagaswamy, J. S. Bergin, and M. C. Wicks, "CoFAR: Cognitive Fully Adaptive Radar," in *IEEE Radar Conference*, Cincinnati, OH, 2014, pp. 0984–0989.

[30] Robey, F. C., S. Coutts, D. Weikle, J. C. McHarg, and K. Cuomo, "MIMO Radar Theory and Experimental Results," in *Conference Record of the Thirty-Eighth Asilomar Conference on Signals, Systems and Computers*, 2004, pp. 300–304.

[31] Rayleigh, L., "On the Law of Reciprocity in Diffuse Reflection," *Phil. Mag.*, Vol. 49, 1900, pp. 324–325.

[32] Manolakis, D., F. Ling, and J. Proakis, "Efficient Time-Recursive Least-Squares Algorithms for Finite-Memory Adaptive Filtering," *IEEE Transactions on Circuits and Systems*, Vol. 34, 1987, pp. 400–408.

[33] Bergin, J. S., J. R. Guerci, R. M. Guerci, and M. Rangaswamy, "MIMO Clutter Discrete Probing for Cognitive Radar," in *IEEE International Radar Conference*, Arlington, VA, 2015, pp. 1666–1670.

4

Introduction to KA Adaptive Radar[1]

4.1 The Need for KA Radar

What does it mean to be adaptive? A working definition that succinctly covers the scope of this book is: "An automated system that adjusts its input-output characteristics on-the-fly to better match its changing environment and mission profile." By environment, in this context, we of course mean the entire radar channel: targets, clutter, jamming, atmosphere, and so forth.

Conventional adaptive radar methods date back to the very beginnings of radar with, for example, automated gain control (AGC) to prevent receiver saturation and cell averaging constant false alarm rate (CACFAR) processing to maintain an acceptable false alarm rate [2, 3]. In the decades that followed, adaptation was extended to antennas (e.g., sidelobe canceller) for jammer

---

1. Portions of this chapter are adapted with permission from [1].

mitigation, and spatio-temporal DoFs for enhanced clutter cancellation (i.e., STAP) [4–6].

Interestingly, virtually all of these adaptive radar subsystems in a modern radar are based on a single adaptation paradigm: "Adjust the input-output characteristics of the subsystem based on the same data stream used to extract signals of interest." In the case of jamming, this is of course essentially the only way to adapt since there is no way to predict when and how a jammer may appear with enough accuracy and precision to form a mitigation filter (e.g., antenna null and/or spectral notch). In contrast, land clutter is quite a different story.

As anyone who has ever attempted to design a STAP filter to process real-world land clutter can attest, it is next to impossible to develop a real-time STAP subsystem (of any appreciable size) that can achieve near ideal performance in all possible real-world geographical/operational environments [1, 7, 8]. The root cause of the issue is the need to estimate on-the-fly (i.e., sampled radar data), the multidimensional statistics (e.g., covariance) required to form a space-time (angle-Doppler) clutter rejection filter. Figure 4.1 depicts several primary causes of so-called heterogeneity/nonstationarity and include [1, 7, 8]:

- Heterogeneous terrain in the form of variations in topology (e.g., mountains) and/or reflectivity (land cover variations,

Figure 4.1 Illustration of some real-world effects that can contribute to so-called clutter inhomogeneity (from [8]).

land-sea interfaces, etc.) as well as internal clutter motion (ICM) such as windblown foliage.
- Large clutter discretes such as man-made buildings, towers, and power lines, for example.
- Road networks that are tantamount to moving clutter (particularly problematic for GMTI radars for obvious reasons).
- Nonlinear/inclined arrays, bistatic/multistatic operation, and so forth, that exacerbate spectral heterogeneity and the interference subspace-leakage (ISL) problem [4].

The above effects, which are generally not mutually exclusive, may result in significant spectral and/or statistical nonstationarities that preclude (or significantly degrade) sample statistics-based adaption methods from achieving their theoretically prescribed performance achievable under i.i.d. conditions [1, 7, 8]. For example, in the case of a STAP subsystem employing N spatial and M temporal DoFs, an $NM \times NM$ interference-only covariance matrix must be estimated (explicitly or implicitly). Even under ideal conditions; that is, the availability of i.i.d. Gaussian samples, an order of $2NM$ samples are required to ensure an output SINR that is, on average, within 3 dB of the theoretical optimum [9] (this can be somewhat relaxed when employing diagonal loading or subspace estimation methods such as principal components, but the number of requisite samples can still be large relative to the stationarity assumption [10, 11]).

Figure 4.2 illustrates one of the most common methods for selecting space-time data samples for estimating the underlying clutter statistics. It is predicated on the assumption that the random clutter samples in adjacent range cells are i.i.d. with respect to that of a cell under test. If complex Gaussianity is further assumed, it can be shown that the sample covariance estimate (matrix formed by averaging the sum of data sample outer products) is in fact the maximum likelihood estimate (MLE) of the true underlying covariance matrix [12]. To prevent the so-called self-nulling problem, and to ensure independence, the cell under test, and several surrounding guard cells may be omitted from the training set (as

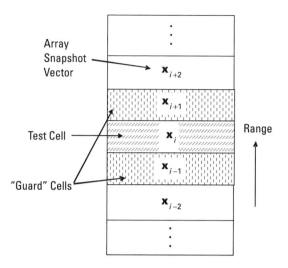

Figure 4.2 Common covariance training data selection strategy.

depicted in Figure 4.2). This step can be relaxed or eliminated altogether if certain precautions are employed, such as diagonal loading or principal components [13].

As an example, the DARPA Mountain Top radar (see Figure 4.3) utilized $N = 14$ spatial and $M = 16$ temporal DoFs in a coherent processing interval (CPI) for a total of $NM = 224$ space-time DoFs, and its range resolution was on the order of about 0.5 km [14]. Thus for full-dimensional STAP (to achieve maximum theoretical performance), statistical stationarity over hundreds of kilometers

Figure 4.3 The DARPA Mountain Top UHF radar experiment [14].

would be required—clearly a very poor assumption over much potential terrain of interest.

While reduced-rank STAP methods can be employed to diminish sample support requirements by as much as an order-of-magnitude, at the expense of generally a few dBs of output SINR [4], stationarity of many kilometers is still necessary. This too is all but precluded in many overland applications (though obviously not all, such as desert terrain).

One might justifiably ask what the impact is of forming sample statistics from nonstationary data. The answer, as it turns out, depends on the nature of the nonstationarity; but in general one or more of the following effects may result [4]:

- Over- or under-nulling of the clutter, particularly in the Doppler regime closest to mainbeam clutter (since this is where clutter is strongest).
- Increased false alarms.
- Target signal loss/radar desensitization.

These effects are particularly acute in GMTI radars that place a premium on detecting so-called slow movers (i.e., minimum detectable velocity (MDV)) [4, 15, 16].

Examples of real-world, and thus generally nonstationary clutter, abound—indeed they are the norm rather than the exception. Figure 4.4 shows a comparison of real clutter data (Figure 4.4(a)) versus that which would have been observed if homogeneous clutter of similar average strength were present (Figure 4.4(b)). Figure 4.5 shows an example of real-world clutter for a higher-resolution experimental X-band system.

In the next section, we introduce the concept of KA processing to help alleviate many of the difficulties encountered when attempting to estimate underlying space-time clutter characteristics over land. While we are primarily focused on radar land clutter, the concept of an environmental database is readily extensible to many other sensors and applications such as bathymetry for shallow water sonar [18] and IR clutter rejection [19].

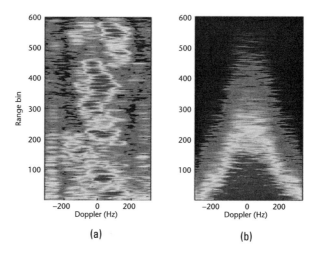

Figure 4.4 Comparison between real-world clutter (a) from the DARPA Mountain Top radar [14], and (b) returns assuming homogenous clutter [1].

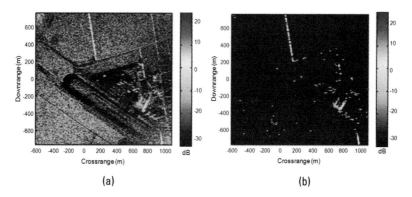

Figure 4.5 Example of real-world X-band radar measurements corresponding to a geographical location with discrete clutter (see Zwycki, et. al. [17] for details): (a) high-resolution geo-registered reflectivity image, and (b) corresponding discrete map.

4.2 Introduction to KA Radar: Back to "Bayes-ics"

The origins of radar KA processing as described herein can be traced to the pioneering work conducted at the Air Force's Research Laboratory (AFRL) at Rome, New York, beginning with expert systems CFAR [20–23], and followed in later years by knowledge-based space-time adaptive processing (KB-STAP) [24–29].

An expert system, from an engineering perspective, attempts to capture the experience and judgment of a human expert in a suitably codified engineered system/algorithm such that to an outside observer the system response to external stimuli mimics that of an expert. This is generally accomplished via some combination of rule-based reasoning [30] and/or adaptive algorithmic structures. The idea behind expert systems CFAR was to improve upon the sophistication of traditional CFAR techniques that typically used some cell averaging approach (mean, median, etc. [31]) based solely on the observed sensor data without the benefit of any exogenous environmental awareness.

KB-STAP extended the above ideas to the multidimensional filtering problem (CFAR is typically a scalar problem). Traditional STAP is at its core a sample covariance-based technique—and thus in essence equivalent to the aforementioned traditional CFAR. KB-STAP research greatly expanded the potential information sources utilized in the overall adaptation process. An early but compelling example of the benefits of adopting a KB-STAP approach was developed by Melvin et al. [28]. By using knowledge of where the road networks (sources of moving clutter) were located in relation to the radar, and a suitable intelligent training strategy (see Chapter 4), the velocity desensitization problem was circumvented [28].

While it is essentially a rhetorical question to ask: "Could my adaptive radar perform better if it had accurate a priori knowledge of the clutter/interference environment?" It is quite another matter to incorporate this prior knowledge—in all its varied forms and quality—into a modern real-time adaptive radar architecture.

Beginning in 2001, the DARPA and AFRL formed a partnership to pursue the KASSPER project (Knowledge-Aided Sensor Signal Processing and Expert Reasoning)—the goal being the development of an entirely new real-time KA-embedded computing architecture that can accommodate the types of KB/KA algorithms that were being developed [32]. While evidence was mounting demonstrating the benefits of enhanced environmental knowledge gleaned from both endogenous and exogenous sources, a fundamental obstacle remained: The knowledge is contained in memory, and thus in general subject to access latencies that can all but preclude its integration into a modern pipelined embedded computing architecture. Section 4.3 below shows how this inherent latency was overcome by a noncausal predictive KA coprocessor architecture.

In this section we will provide an overview of KA radar, from algorithms to a high-level overview of real-time high-performance embedded computing (HPEC). As the title of this section implies, the use of prior knowledge is certainly not new in signal processing or statistical analysis. Indeed the well celebrated Bayesian approaches are predicated on statistical priors [12]. However real-world prior information comes in such a myriad of forms, and not usually as a probability density function as prescribed by Bayes, that a variety of approaches must be adopted to fully exploit it [1, 8].

Algorithmically KA processing in general, and KA-STAP in particular, can be delineated into so-called *indirect* and *direct* methods of incorporating prior information. In Section 4.2.1, we survey indirect methods such as intelligent training and filter selection, while in section 4.2.2 we consider some examples of direct methods such as Bayesian filtering and data prewhitening.

This is a very dynamic and diverse field that is rapidly evolving at the time of this writing. Thus, the treatise herein is by no means complete—but rather is aimed at providing the reader with a working sense of the possibilities. Moreover, our focus here is on the STAP/CFAR problem, and not on, for example, KA tracking—which is a rapidly developing subfield in its own right [33].

4.2.1 Indirect KA Radar: Intelligent Training and Filter Selection

In the Intelligent Training and Filter Selection (ITFS) approach, prior knowledge of the interference environment is used to optimize two adaptive filtering processes: (1) the filter selection; and (2) the selection of so-called filter training data (i.e., the selection of data samples used to form a sample covariance estimate—either directly or implicitly (e.g., data domain methods) [34]. In the case of radar clutter, this could be accomplished by first conducting an environmental segmentation analysis based on whatever prior terrain/clutter database is available. Everything from digital terrain and elevation data (DTED) to land cover land use (LCLU) to synthetic aperture radar (SAR) imagery, or even hyperspectral imagery can be used [1]. Land clutter tends to be clumpy; that is, it tends to be locally similar—but with distinct and often abrupt boundaries (see Figure 4.6 for example). Clearly from physical principles, an adaptive filter should not attempt to lump all these regions together and apply a single filtering strategy. Instead, a segmentation analysis should be performed and an adaptive filter tailored to that region should be applied.

Generally speaking, the filter selection stage determines what type of adaptive filter is best suited to a given segmented region. In the case of STAP filtering for clutter suppression in radar, a pivotal step is the domain in which the actual filtering is performed (e.g., pre- or post-Doppler element or beamspace [4, 6, 35]) and the number of adaptive degrees-of-freedom (ADoFs)—which manifests itself ultimately in the size of the adaptive filter (e.g., the size of the sample covariance matrix employed [4]). For example, in the case of the principal components (PC) method, the number of adaptive DoFs refers to the number of significant eigenvectors to be included in the adaptive weight calculation [13]. Similarly for the multistage Weiner filter (MWF), the number of stages is the metric for AdoFs [36]. What is critical is that the number of ADoFs be matched to the available training data (and of course the real-time computing architecture). A general rule of thumb, which has its origins in the RMB result [9], is that there be an order

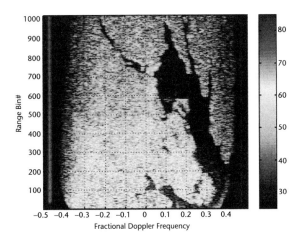

Figure 4.6 Measured range–Doppler returns for the MCARM data set showing the highly segmented nature of radar clutter returns [17].

k to $2k$ i.i.d. samples available for training the adaptive weights (e.g., sample covariance estimation), where k is the number of effective AdoFs [4]. This result is predicated on the assumption that the k dominant eigenvectors are strong relative to the noise floor. Indeed for strong enough interference, k linearly independent samples (snapshots) completely defines the subspace—which is the basis for the Hung-Turner projection method [37].

Once a basic filtering structure has been selected, a training strategy can be selected and optimized for that choice. Basically, all or a subset of the samples from the locally stationary region are utilized in the weight training stage. In the case of PC, all of the range bins—including the cell under test—might be included since it has been shown that this approach is robust to target cancellation. In contrast, a multibin post-Doppler approach might need to take extra care and introduce exclusion and guard cells to prevent target signal cancellation [6].

Figure 4.7 illustrates the impact that ITFS can have when applied to real-world data. As described in [28], the MCARM data set included a number of significant highways—that is, moving clutter. If one simply applied traditional sample averaging techniques such as those previously described, one could suffer significant detection losses at roadway speeds [28]. Using an intelligent training and adaptation scheme, which essentially took account of the road networks, a significant improvement in detection was achieved.

Example 4.1 Intelligent Filter Selection: Matching the Adaptive DoFs (ADoFs) to the Available Training Data

Consider an airborne, side-looking MTI radar employing a ULA comprised of $N = 8$ receive antenna elements with half-wavelength spacing, and $M = 16$ pulses available for processing in a CPI (see [4] for extensive analyses of such systems). Theoretically, a full DOF STAP processor would utilize $NM = 128$ ADoFs that, to first order, would require approximately 100 to 200 samples (i.e., range bins) or more depending on the level of diagonal loading employed (assuming a sample matrix inverse (SMI) approach [4]). It is not out of the question that this range extent might correspond with many kilometers or evens tens of kilometers.

4.2 Introduction to KA Radar: Back to "Bayes-ics"

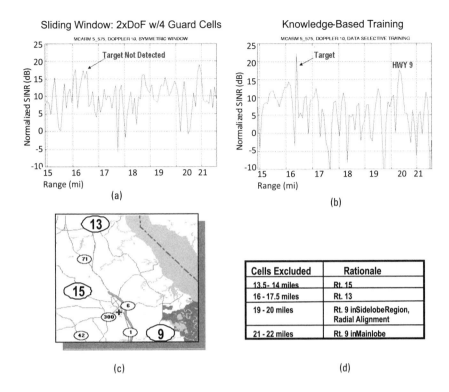

Figure 4.7 Illustration of the impact prior knowledge (in this case prior road network data) can have on improving detection performance for the MCARM data set. (a) STAP filter residue without knowledge-aided processing. (b) Target, which was previously undetected, is clearly visible after intelligent training (see [28] for further details). (c) Local map of region indicating locations of road networks. (d) A training cell exclusion rule based on the map data.

At the heart of KA processing is a dynamic environmental database which, among many possible uses, can be exploited to determine the relative degree of stationarity (e.g., clutter homogeneity) over a given region [1, 8]. Unless one happens to be flying over flat desert or grasslands, it is quite often the case that land clutter is comprised of a variety of regions (e.g., roads, buildings, fields, mountains, and lakes) [38]. Methods for determining the degree of stationarity in a sample data set abound (see for example [28, 39] and the references cited therein related to the nonhomogeneity detector approach (NHD) that has been successfully applied to land clutter).

Assume that we wish to operate our radar (at maximum STAP efficiency) over a region for which quasi-stationarity may only be assumed for about 50-range bins. Applying the "twice the number of ADoFs rule for sample support" [4, 6, 11, 35] yields a maximum number of 25 for the allowable ADoFs (rank of the STAP filter).

Fortunately, there is a plethora of so-called reduced-rank STAP methods that have been developed over several decades that can be utilized to transform the full DoF STAP filter into a reduced-rank one that meets the ADoFs constraint—while often maintaining acceptable SINR loss (see [4, 6, 35] and references cited therein).

Table 4.1 provides several examples for which the product of the space and time ADoFs is less than (or equal to) 25. The top portion of the table is a sampling of data-independent rank-reduction methods inasmuch as the number of spatial and temporal ADoFs does not directly depend on the statistical properties of the received data—in contrast to the lower portion of the table comprised of data-dependent methods for which the rank, k, will depend on the data (e.g., number of principal components and/or MWF stages [36]). For these latter methods, an estimate for k (interference subspace rank) can be obtained from Brennan's rule, which states that the rank of the clutter subspace is approximately given by

Table 4.1
A Sampling of Various Reduced-Rank STAP Methods that Meet the Sample Support Constraint of 50-Range Bins but Also Attempts to Maximize the Number of ADoFs*

Reduced-Rank Technique	Spatial DoFs	Temporal DoFs	Space-Time ADoFs
Post-Doppler element space	8	3^1	24
Post-Doppler beamspace	3	8^1	24
PRI staggered element space	8	3^2	24
PRI staggered beamspace	3	8^2	24
ADPCA	8	3^3	24
Principal components (PC)	N/A	N/A	k^4
Multistage Wiener filter (MWF)	N/A	N/A	k^4

*See [4, 6, 35] for further details
1 Number of Doppler bins utilized.
2 Number of PRI Doppler staggers.
3 Number of pulses processed jointly and adaptively.
4 Rank of clutter subspace.

$$k \approx \lceil N + \beta(M-1) \rceil \qquad (4.1)$$

where $\lceil \; \rceil$ denotes the ceiling operator (i.e., round-up to the nearest integer), and β is a measure of the antenna motion relative to the PRI (i.e., distance, measured in half-wavelengths, traversed per PRI [6]). For the case at hand β would need to satisfy

$$\beta \leq \frac{25-N}{M-1} = \frac{17}{15} \qquad (4.2)$$

to ensure that enough sample support is available for training purposes.

In practice, k obtained from (4.1) is a bit optimistic, that is, the actual clutter subspace rank is generally greater due to a multitude of real-world effects (e.g., internal clutter motion (ICM), antenna crabbing, and jitter [4]). Also, the rank for the MWF (measured in the number of stages utilized) is generally bounded by (4.1)—but may often be less due to its steering vector dependency [40].

4.2.2 Direct KA Radar: Bayesian Filtering and Data Prewhitening

In the Bayesian approach to KA radar STAP, prior knowledge is incorporated directly by the filter to aid in adapting to nonstationary clutter. A convenient pedagogical framework for this approach is the Bayesian covariance estimation approach described by T. W. Anderson [12].

Wishart [41] established that the elements of a sample covariance matrix $[L\hat{\mathbf{R}}]_{i,j}$ formed from an outer product sum of L Gaussian i.i.d. samples,

$$\hat{\mathbf{R}} = \frac{1}{L}\sum_{i=1}^{L} x_i x_i' \qquad (4.3)$$

obey a Wishart distribution (actually a complex Wishart for the case at hand [42]) of degree L, that is, $\hat{\mathbf{R}} \sim W(L\hat{\mathbf{R}}, L)$. Fortunately, (4.3) also corresponds to the maximum likelihood estimate of the underlying covariance matrix $\mathbf{R}$ [12, 42].

If a prior estimate of the covariance matrix exists, say $\hat{\mathbf{R}}_0$, it is not unreasonable to assume it too is similarly Wishart distributed—particularly if it was formed over the same geographical region. Moreover, if it is based on prior radar observations, then it is also of the form in (4.3); that is, a sum of data-sample outer products. The corresponding Bayesian (maximum a posteriori) estimate that combines $\hat{\mathbf{R}}$ and $\hat{\mathbf{R}}_0$ is easily derived: Let L_1 and L_0 denote the degrees of $\hat{\mathbf{R}}$ and $\hat{\mathbf{R}}_0$, respectively, which are further assumed to be i.i.d. and complex Wishart [12]. Then $\hat{\mathbf{R}}$ and $\hat{\mathbf{R}}_0$ are collectively sufficient statistics for $L_0 + L_1$ i.i.d. samples $\{x_i: i = 1, ¼, L_0 + L_1\}$. Thus, the maximum a posteriori solution of $\hat{\mathbf{R}}$ given prior $\hat{\mathbf{R}}_0$ is equivalent to the maximum likelihood solution based on all of the data $\{x_i: i = 1, ¼, L_0 + L_1\}$, [12]

$$\begin{aligned}\hat{\mathbf{R}} &= \max_{\mathbf{R}} f(x_i : i = 1, \ldots, L_1 | \mathbf{R}) f_0(\mathbf{R}) \\ &= \max_{\mathbf{R}} f(x_i : i = 1, \ldots, L_0 + L_1 | \mathbf{R}) f_0(\mathbf{R}) \\ &= \frac{1}{L_0 + L_1}\left(L_0 \hat{\mathbf{R}}_0 + L_1 \hat{\mathbf{R}}_1\right)\end{aligned} \quad (4.4)$$

where: $f_0(\mathbf{R})$ denotes the prior pdf associated with the prior covariance estimate $\hat{\mathbf{R}}_0$ based on L_0 samples—and thus is $W(L_0 \hat{\mathbf{R}}_0, L_0)$; and $\hat{\mathbf{R}}_1$ denotes the maximum likelihood (ML) estimate based on samples. Equation (4.4) has an obvious intuitive appeal: the a posteriori covariance estimate is formed as a weighted sum of the prior and current estimates with weighting factors proportional to the amount of data used in the formation of the respective sample covariances.

An obvious yet useful generalization of (4.4) is

$$\begin{aligned}\hat{\mathbf{R}} &= \alpha \hat{\mathbf{R}}_0 + \beta \hat{\mathbf{R}}_1 \\ \alpha + \beta &= 1\end{aligned} \quad (4.5)$$

which is the familiar colored loading or blending approach of Hiemstra [43] and Bergin et al. [44], respectively. The practical advantages of (4.5) relative to (4.4) are many. For example, the data used to form the prior covariance might lose its relevance with time—the so-called stale weights problem [45]. In that case, even though $\hat{\mathbf{R}}_0$ might have been formed from L_0 samples, it effectively has less information and should be commensurately deweighted. A common method for accomplishing this, borrowed from adaptive Kalman filtering, is the fading memory approach in which case α, in (4.5), is

$$\alpha = e^{-\gamma t} L_0 \qquad (4.6)$$

where t is the time elapsed since the prior covariance estimate was formed, and the positive scalar γ is the fading memory constant [46].

In a more general setting, the blending parameters (α, β) could be chosen based on the relative confidence in the estimates. For example, $\hat{\mathbf{R}}_0$ could be derived from a physical scattering model of the terrain. In which case it is also typically of the form (4.3) with the distinction that the outer products represent clutter patch steering vectors weighted by the estimated reflectivity [4, 47],

$$\hat{\mathbf{R}}_0 = \frac{1}{N_c} \sum_{i=1}^{N_c} G_i v_i v_i' \qquad (4.7)$$

N_c clutter patches have been utilized in the formation of $\hat{\mathbf{R}}_0$ (typically corresponding to a particular iso-range ring [4, 47]) where $v_i \in C^{NM}$ is the space-time (angle-Doppler) steering vector corresponding to the ith clutter patch and G_i its corresponding power [4]. Such information could be available a priori from SAR imagery [1, 8] (essentially a high-resolution clutter reflectivity map) or physics-based models [48].

Though the confidence metric to apply, in the form of the weighting parameter α, is difficult to ascribe in practice since the quality of the a priori estimate is vulnerable to a number of error sources, a straightforward remedy is to choose a adaptively so as to maximally whiten the observed interference data [49]. For example:

$$\min_{\{\alpha\}} Z_L(\alpha) \qquad (4.8)$$

where

$$Z_L(\alpha) = \left\| \sum_i y_i y_i' - I \right\| \qquad (4.9)$$

and where

$$y_i = \left(\alpha \hat{\mathbf{R}}_0 + \beta \hat{\mathbf{R}}_1\right)^{-\frac{1}{2}} x_i \qquad (4.10)$$

In (4.8) to (4.10), x_i is the space-time snapshot vector for the ith range bin; $(\alpha \hat{\mathbf{R}}_0 + \beta \hat{\mathbf{R}}_1)^{-\frac{1}{2}}$ is the whitening matrix corresponding to a particular a; y_i is the vector residue with $\dim(y_i) = \dim(x_i)$; and the summation in (4.9) is performed over a suitable subset of the radar observations for which $\hat{\mathbf{R}}_0$ is believed valid. If an a priori covariance estimate is available for each range bin, then the vector residue can be replaced with

$$y_i = \left(\alpha \hat{\mathbf{R}}_0(i) + \beta \hat{\mathbf{R}}_1\right)^{-\frac{1}{2}} x_i \qquad (4.11)$$

where $\hat{\mathbf{R}}_0(i)$ is the a priori estimate for the i-th range bin.

The above adaptive a approach is but a special case of an entire class of direct filtering methods incorporating prior information, viz., *data prewhitening* (or simply data detrending). In a more general setting, the space-time vector residues, $\{y_i\}$, can be viewed as a detrended vector time series using prior knowledge in the form of a covariance-based whitening filter. The major potential advantage of this is to remove (or attenuate) the major quasi-deterministic trends in the data (e.g., clutter discretes and mountains) so that the resulting residue vector time series is less nonstationary or inhomogeneous.

An interesting example of this can be found in [17]. In this prewhitening example, a CLEAN algorithm was applied to the APTI data set of Figure 4.7(a), resulting in the discrete map of Figure 4.7(b). A deterministic covariance [47] was then formed as in (4.7), from which a square root whitening filter matrix could be derived. Figure 4.8 shows a log exceedance plot of the difference between the unwhitened data and the prewhitened data. Note the presence

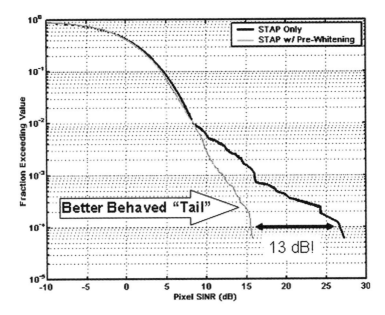

Figure 4.8 Results of applying the prewhitening approach. Note the significant reduction in the so-called clutter tails in the log exceedance plot [17].

of spiky clutter as evidenced by the so-called fat-tails in the unwhitened data, and its subsequent reduction when employing direct KA methods.

Example 4.2 Using Past Observations as a Prior Knowledge Source

Perhaps the simplest form of a knowledge environmental database can be created from previous scans over a given geographical region. Indeed the preponderance of multipass SAR systems, including coherent change detection (CCD) methods [50], provides an extremely rich knowledge source for an environmental clutter database [15]. Inevitably, however, there will likely be some amount of decorrelation between the prior measurements and the current received radar signals due to a variety of factors including, but by no means limited to, georegistration errors, sensor orientation differences, internal clutter motion, and so forth.

A convenient mechanism for modeling vector random variable decorrelation is the covariance matrix tapering (CMT) approach [45]. Specifically, an NM dimensional complex multivariate RV with associated original covariance $R \in C^{NM \times NM}$, that has undergone a decorrelating process [45], can simply be described as

$$\hat{R}_0 = R \circ T \in C^{NM \times NM} \quad (4.12)$$

where $\hat{R}_0$ is the covariance matrix resulting from a decorrelating process characterized by the CMT $T \in C^{NM \times NM}$, which in general is a positive semidefinite (and conformal) matrix [45]. From the Schur product theorem, $\hat{R}_0$ is guaranteed to be positive definite if R is positive definite and T is at least positive semidefinite [51].

Due to real-world clutter nonstationarity and the desire to minimize the sample support required to estimate the requisite sample covariance matrix (implicitly or explicitly), we would like to take advantage of the prior information in the form of a decorrelated estimate of the true underlying covariance; that is, $\hat{R}_0$ from (4.12). Let us assume that the primary source of decorrelation is due to range walk, which to first-order could be caused by a slight misalignment of the sensor range bin grid [4]. In this case, the CMT is of the form

$$T = T_{time} \otimes \mathbf{1}_{N \times N} \quad (4.13)$$

where denotes the matrix Kronecker product [51], $\mathbf{1}_{N \times N}$ denotes the Hadamard identity matrix (unity elements), and the temporal CMT, $T_{time} \in C^{M \times M}$, is given by

$$T_{time} = \begin{bmatrix} 1 & \rho & \cdots & \rho \\ \rho & 1 & & \vdots \\ \vdots & & \ddots & \rho \\ \rho & \cdots & \rho & 1 \end{bmatrix} \quad (4.14)$$

where the correlation coefficient, $0 << \rho << 1$, captures the degree of decorrelation between the current and prior data.

If we denote the current and prior sample covariance estimates by $\hat{R}_0$ and $\hat{R}$, respectively, we would like to form a Bayesian-like estimate along the lines of (4.4) and (4.4),

$$\hat{\mathbf{R}}_{KA} = \alpha\hat{\mathbf{R}}_0 + (1-\alpha)\hat{\mathbf{R}} \qquad (4.15)$$

where $0 \leq \alpha \leq 1$ is a suitably chosen blending parameter. If there was no decorrelation between the prior data and the current measurements, they represented i.i.d. samples from the same underlying Gaussian (clutter) distribution, then one could invoke the exact Bayesian solution of (4.4). However, due to real-world decorrelating effects, the effective amount of prior data will generally be less than L_0, leading to the inequality

$$\alpha \leq \frac{L_0}{L_0 + L_1} \qquad (4.16)$$

where L_1 denotes the number of i.i.d. samples used to form the current sample covariance, per (4.4).

For the special case where $L_0 = L_1$, (4.16) implies that in general $\alpha \leq \frac{1}{2}$, unless (of course) ρ in (4.14) is unity (i.e., no decorrelation), in which case α should be set equal to 0.5 for the optimal Bayes' estimate.

An interesting question is thus: Given the amount of decorrelation as described by the CMT of (4.14), how can one choose the effective number of samples, and thus α? An approximate answer can be obtained by computing relative SINR loss versus ρ in and (4.14) equating it with the Reed, Mallet, Brennan (RMB) rule [9], which is given by

$$\mathrm{SINR}_{\mathrm{Loss}} = \frac{L - NM + 2}{L + 1}, \quad L \geq NM \qquad (4.17)$$

where L denotes the number of i.i.d. complex Gaussian samples. For example, if r in (4.14) results in a 5-dB loss in predicted SINR (relative to optimum), it is (from (4.17)) effectively equivalent to a sample covariance comprised of approximately 1.33 NM samples (assuming NM is relatively large). Assuming that $2NM$ samples were originally used to form $\hat{\mathbf{R}}_0$, we see that due to decorrelating effects, the number of effective samples (effective weighting) has been discounted by about 34%.

To illustrate the above observations, consider an airborne side-looking MTI radar utilizing an $N = 16$ ULA (half-wavelength element spacing operating in the far field), and $M = 16$ pulses in a coherent processing interval (CPI). Assume further that $b = 1$, and that the CNR is 40 dB. Plotted in Figure 4.9 is relative SINR loss versus Doppler for the ideal (exact covariance case), the case when the prior covariance is utilized but with a decorrelation due to range walk corresponding to $r = 0.80$, (corresponds to about a 5–7-dB SINR loss), the case when only NM samples of the current data is used (no priors) resulting in nearly a 30-dB loss (no diagonal loading), and finally the case when the current estimate is blended with the prior estimate that has been discounted based on (4.17). Note that suitably blending the prior estimate, even though it has undergone decorrelation, can still yield a better estimate of the current covariance matrix—illustrating one of the basic benefits of KA processing.

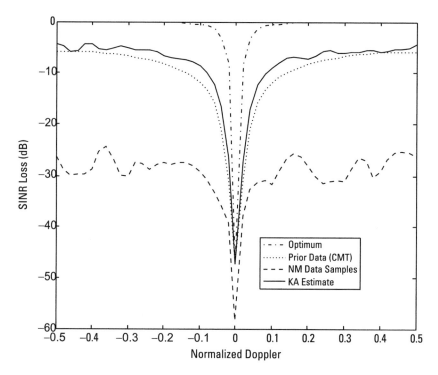

Figure 4.9 Example of the benefits of using prior covariance estimates from previous looks at a certain geographical location.

In the next section, we address the seemingly daunting challenge of incorporating prior knowledge—an inherently memory intensive process—into a high-performance embedded computer.

4.3 Real-Time KA Radar: The DARPA KASSPER Project

Over the years, ingenious real-time computing architecture solutions have been devised to implement the sample matrix-based (maximum likelihood) STAP solutions in real-time flight hardware [34]. In particular, to achieve the enormous throughput burden of a modern multichannel STAP radar, highly parallel HPEC systems based on so-called data domain reformulations of the Wiener-Hopf equation have been devised [8, 9]. Figure 4.10 shows one such architecture based on a QR-factorization solution to the Wiener-Hopf equations [52].

With such architectures, tens to hundreds of GFLOPs (billions floating point operations) of real-time computing power can be achieved in hardware that can fit on an airborne radar aircraft [53]. Though marvels of modern technology, these machines are

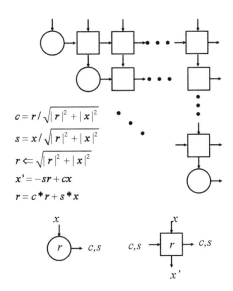

Figure 4.10 An example of computer array implementing a data domain reformulation of the sample matrix-based Wiener-Hopf equation (see for example [52]).

cyber savants: they can solve the Wiener-Hopf equations at blinding speeds in a strict pipelined fashion but grind to a snail's pace if the data flow is disrupted for nonpipelined operations. This is a major fundamental obstacle to implementing knowledge-aided or generalized Bayesian approaches which are inherently memory intensive (prior information needs to be stored) [1]. Figure 4.11 shows the order-of-magnitude time scales for accessing different memory storage devices. Thus, to create a real-time knowledge-aided HPEC (KA-HPEC, pronounced "K-PEC") architecture, a major breakthrough in memory management must be achieved since much of the prior information (e.g., terrain maps, road networks, and discrete maps) will reside on mass storage (and thus slow) media.

4.3.1 Solution: Look-Ahead Scheduling

The key KA-HPEC breakthrough in the DARPA KASSPER project is based on a basic fundamental insight:

> There is a significant degree of determinism—and thus predictability—to radar clutter returns, particularly if the prediction horizon is only on the order of seconds.

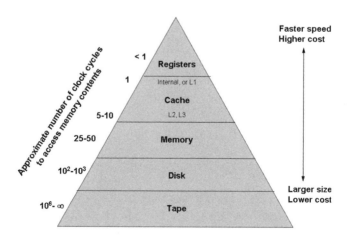

Figure 4.11 Illustration of the time scales involved in accessing different memory storage media (Source: Dr. D. Patterson, graduate computer architecture course, University of California, Berkeley, Spring, 2001).

For example, let t_0 denote the present time of a postulated airborne radar. Let $t_0 + \delta t$ denote a time slightly in the future—say $\delta t = 1$ second. Then in practice, the following are true:

1. The location of the aircraft at $t_0 + \delta t$ can be predicted to a very high degree of accuracy assuming that no radical maneuvering is occurring.

2. The future state of the radar (e.g., look-direction, frequency, PRF, etc.) at $t_0 + \delta t$ is also known to a very high degree of accuracy.

The justification for the first assertion is simply that given the full kinematic state vector of the aircraft (position, speed, heading, etc.), Newtonian mechanics insures fairly deterministic behavior— particularly for just a few seconds into the future. Justification for the second assertion arises from the simple fact that modern airborne radar systems utilize a radar scheduler [54]. Since the radar is computer controlled, it must have a tasking schedule. The scheduler is highly deterministic when considering a future time horizon on the order of seconds.

Why are the above assertions so critical to solving the posed memory access problem? Simple: they allow for look-ahead scheduling. More specifically, they allow for *noncausal* processing whose prediction horizon is commensurate with the memory access delays! To see how this can be exploited by a KA-HPEC architecture, consider Figure 4.12. In this instantiation, a noncausal look-ahead computer is running in parallel with a conventional causal STAP HPEC processor. The noncausal processor is used to spot trouble before it occurs, and to perform the necessary memory retrieval and precomputations to ensure that the right weight modification scheme is ready to go when the data appear.

Figure 4.13 shows the MIT Lincoln Laboratory KASSPER HPEC system, a real-time 96-node parallel processing architecture implementing the noncausal look-ahead scheduling scheme of Figure 4.12 [27]. The system has the capability of receiving real-time I & Q (in-phase and quadrature) digitized samples from multiple receive channels over the full range extent of a radar and imple-

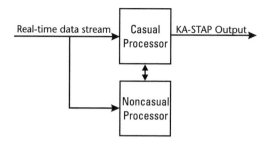

Figure 4.12 Example of a KA-HPEC architecture exploiting the high degree of radar determinism with look-ahead time scales on the order of seconds. The noncausal processor, running in parallel with a more conventional HPEC STAP processor, is used to look-ahead and detect regions of the radar field of regard requiring knowledge-aided processing—and thus modifications to the normal adaptive weights calculations.

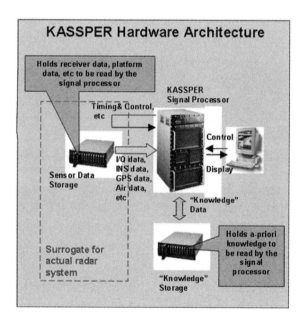

Figure 4.13 The MIT Lincoln Laboratory 96-node real-time KASSPER HPEC system.

menting a variety of knowledge-aided algorithms throughout the entire radar signal processing chain from STAP to CFAR.

Though when it comes to real-time HPEC the devil is most certainly in the details, Figure 4.14 gives the basic gist of how the look-ahead scheduling is implemented. As the aircraft moves, a sliding window of data is migrated from a mass storage medium (e.g., disk drives) to a more readily accessible location (e.g.,

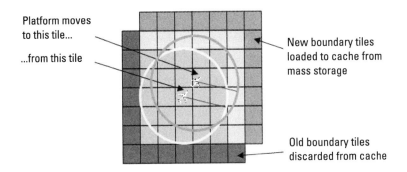

Figure 4.14 Illustration of the environmental database manipulation illustrating the sliding window approach to migrating data from mass storage to RAM and ultimately to cache.

RAM). Depending on the particulars of the radar tasking, a first-pass decision is made as to what regions require knowledge-aided processing. For example, if the radar is scheduled to point in a direction where a major road network is known to exist, essential details regarding this road network (orientation, range extent, etc.) are extracted and exploited in potentially several stages of the radar signal processing chain. Given the look-ahead time buffer, this is all accomplished prior to the actual radar event. The exact extent of the sliding window depicted in Figure 4.14 depends (of course) on the particular radar parameters (min/max range, altitude, etc.).

Example 4.3 Balancing Throughput in a KASSPER HPEC Architecture

Assume it is desired to upgrade a conventional STAP architecture capable of, say, 100 GFLOPs (pipelined) throughput to a KA-STAP configuration. However, due to both budgetary and form-factor constraints, only an additional 10 GFLOPs of real-time processing and 4 GB of high-performance RAM can be added to the front-end. We will assume that the addition of mass storage (e.g., RAID) is not a problem as it can reside elsewhere on the platform. This latter assumption allows for a very large dynamic environmental database containing both a priori data and data gleaned during flight.

As shown in Figure 4.15, this upgrade is tantamount to introducing an environmental dynamic database (EDDB)—subsystem

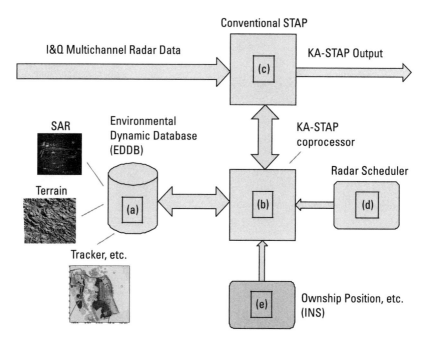

Figure 4.15 Illustration of a high-level canonical KA-STAP HPEC architecture.

(a), a KA coprocessor (b) that communicates with the database (bidirectionally), the conventional STAP processor (c), the radar scheduler (d), and the ownship location system (e) (e.g., INS). Basically, the KA coprocessor (b) looks ahead by determining where the radar will be (from (e)), and where it will be pointing as well as what it will be transmitting, all information available from the scheduler (d). The coprocessor (b) then scans through the EDDB (a) identifying potential problem areas and computing the requisite modifications (e.g., training data segmentation, and colored loading).

To make this example a bit more definitive, we will further assume that the bulk of the computational burden in the KA coprocessor (b) is assigned to computing statistics such as the sample covariance matrix—somewhat similar to the types of computations performed in the conventional STAP processor (c). Clearly, since (b) has only one-tenth the compute power of (c), it could not possibly process all of the data in the EDDB. This leads naturally to one of the key features of the EDDB: database element tags that

4.3 Real-Time KA Radar: The DARPA KASSPER Project

allow for quick scanning and identification of potential problem areas. These elemental features could include reflectivity, type of surface (including presence or absence of buildings and/or roads, etc.), topology, and so forth. The point being that a quick scan by the KA coprocessor (c) could identify and rank-order potential problem areas—then compute relevant statistical information to be used in the conventional STAP processor (b) [1].

Let DR denote the range extent that the conventional STAP processor (b) can process in real-time. This naturally leads to the first throughput constraint on the KA coprocessor (c) when, as we have assumed, the computations are of a similar type to (b), viz.,

$$\Delta R_{KA} \leq 0.1 \Delta R \qquad (4.18)$$

where ΔR_{KA} denotes the amount of range extent that the KA coprocessor can process in real-time. Equation (4.18) is merely a reflection of the obvious fact that if (c) has only one-tenth the compute power of (b), then to first-order it can only process at most one-tenth the amount of data as (b).

The next design issue is how far into the future does the KA coprocessor need to look? The answer to this question is driven by the latency associated with loading the results from the KA coprocessor (c) into the conventional STAP processor (b). Though a precise answer to this question is inextricably dependent on the details of the HPEC system, a look-ahead on the order of seconds is typically more than sufficient [1]. Fortunately this amount of look-ahead is very easily achieved since both ownship future location and radar schedule are all but deterministic on these time scales. For simplicity, we will assume a look-ahead requirement of 1 second.

If each element of the EDDB corresponds to a 1 m² patch of ground, and the range of extent for the radar is 200 km, with min-max ranges (r_{min}, r_{max}) of 10 to 210 km, respectively, then for each pulse and/or CPI the amount of area A to be examined by the KA coprocessor is on the order of

$$A \approx 4\pi \left(r_{max}^2 - r_{min}^2 \right) = 263 \text{ million m}^2 \qquad (4.19)$$

which, in turn, implies that on order of 263 million database elements need to be examined on a continuous (real-time) basis.

Assuming that each database element consists of 128 bits (reflectivity, elevation, cover type, etc.), the above corresponds to approximately 34 GB of storage. Since this data would need to reside in fast-access RAM to be processed by the KA coprocessor, we see that there is a problem since we have only been allocated 4 GB of RAM in the upgrade.

The simplest and practical remedy is to rank order each element offline or in the background based on known deleterious characteristics. For example, EDDB elements associated with large discretes of sufficient reflectivity to cause false alarms would have a high priority, while elements embedded in a homogenous background would rank low or be culled altogether. Thus, to first-order, 4/34 or approximately 12% of the land clutter could be processed using KA methods—in one form or another. Since in an operational setting there may often be hot spots or areas of particular interest, 12% might be sufficient.

4.3.2 Examples of a KA Architectures Developed by the DARPA/AFRL KASSPER Project[2]

A number of researchers participated in the DARPA/AFRL KASSPER project and developed a number of successful KA methods and architectures, and the interested reader is encouraged to consult the literature and references cited therein, particularly the proceedings of the KASSPER workshops [32] for further details. While a number of KA algorithms and architectures have been developed and demonstrated, we will highlight the particular KA STAP instantiation developed by Information Systems Laboratories (ISL), Inc. [55], and the Georgia Institute of Technology [7].

Figure 4.16 shows a high-level component diagram of ISL's KA-STAP (IKA-STAP) architecture adopted from [55]. In addition to the EDDB (upper left-hand corner of figure), the architecture is seen to consist of a KA coprocessor (top shaded region) and the conventional STAP processing chain (bottom shaded region). Note

---

2. The author is indebted to Jamie Bergin and Paul Techau, ISL, Inc., and William Melvin et al., GTRI, the developers of the architectures described in this section.

4.3 Real-Time KA Radar: The DARPA KASSPER Project

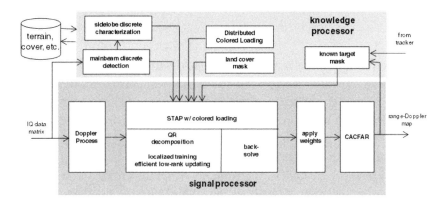

Figure 4.16 Illustration of the IKA-STAP architecture developed by Information Systems Laboratories, Inc., under the DARPA/AFRL KASSPER project.

that the KA coprocessor communicates with the target tracker to facilitate various blanking operations. Not shown are communication paths with the radar scheduler and ownship INS system.

As evidenced by the KA coprocessor delineation into mainlobe/sidelobe discretes and distributed clutter, the IKA-STAP architecture incorporates special processing for optimum clutter discrete handling, as well as distributed clutter cancellation [55]. In particular, five distinct clutter mitigation elements are incorporated [55]: (1) colored loading for distributed clutter, (2) colored loading for main-beam discrete clutter, (3) colored loading for sidelobe discrete clutter, (4) land cover-based training data editing, and (5) iterative target editing.

Colored loading [43], as the name implies, is an additive blending of sample covariance estimates derived from both online and KA sources. In particular, the colored loading covariance estimate employed in the IKA-STAP architecture is of the form [55],

$$\hat{\mathbf{R}}_{KA-CL} = \hat{\mathbf{R}}_{SM} + \beta_{KA}\hat{\mathbf{R}}_{KA} + \beta_{DL}I \qquad (4.20)$$

where $\hat{\mathbf{R}}_{SM}$ is a sample covariance based estimate derived from the online measurement data, $\hat{\mathbf{R}}_{KA}$ is a purely knowledge-based estimate, and (b_{KA}, b_{DL}) are positive scalar blending parameters associated with the KA covariance estimate and conventional diagonal loading. Note that but for the additional diagonal loading term,

(4.20) is of the form found in (4.5). It should be noted that while we are using a power domain or direct covariance representation approach to simplify the analysis, data domain representations employing QR factorizations are readily implementable. See for example Chapter 6 in [4], which shows how to append additional additive (and even Hadamard multiplicative) terms to the STAP data cube used for sample covariance estimation.

For the GMTI application considered in the IKA-STAP architecture development, the main culprits contributing to deleterious conventional STAP performance were main lobe and sidelobe discretes and abrupt clutter region changes (boundary interfaces). Once these effects were identified and accounted for, the remaining whitened clutter statistics were fairly benign—and thus amenable to conventional sample covariance and diagonal loading methods [4].

In the case of mainbeam discretes, which by definition have high signal strength relative to noise, it is possible to directly identify and estimate requisite color loading parameters [55]—as illustrated in Figure 4.17. It consists of forming a quasi-SAR (range-Doppler) map matched, thresholding the image to allow only large returns around zero Doppler then forming augmentation steering vectors to the colored loading covariance. Note that the both tracker data and a priori EDDB entries aid in ensuring that real targets are not misidentified and that a discrete is declared with high confidence. The result is a KA covariance that has deeper nulls than would otherwise be the case.

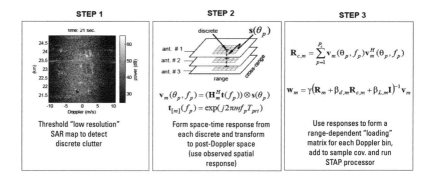

Figure 4.17 Illustration of the method employed to detect and characterize main-lobe clutter discretes and incorporate KA corrections via colored loading [55].

Figure 4.18 shows the subsystem for addressing large sidelobe discretes that can show up as false targets. Both a priori knowledge from a variety of environmental databases, along with the data gathered from the mainbeam scheme described above, provide a rich basis for predicting when and where (range-Doppler-angle) a sidelobe discrete is likely to arise. While sidelobe blanking methods could be readily implemented [52], a more sophisticated technique was implemented in IKA-STAP wherein a suitable preprocessed snapshot vector known to contain the sidelobe discrete is added to the covariance estimate via colored loading (see [55] for details). This allows for detection of targets (unlike blanking) in the range swath known to be vulnerable to sidelobe clutter discretes. Interestingly, the null-on-transmit technique described in Chapter 2 is potentially an even more effective technique as the offending clutter discrete is effectively removed from the receiver processing chain.

Also contained in the IKA-STAP architecture is the provision for land cover-based training data editing—that is the use of terrain maps to determine regions of relative homogeneity, abrupt boundaries, and so forth. Figure 4.19 illustrates how such maps can be projected into the radar's field of view to determine how best to segment the training data (see [55] for further details).

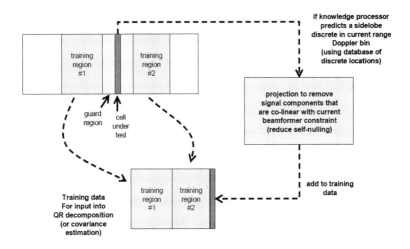

Figure 4.18 IKA-STAP subsystem for addressing the sidelobe clutter discretes problem [55].

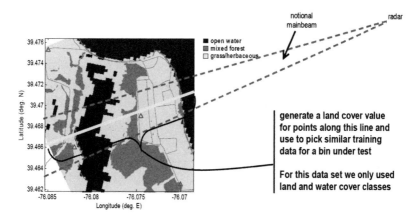

Figure 4.19 IKA-STAP subsystem for handling the intelligent training data selection problem [55]. In this example, land cover maps are projected into the radar's field of view to determine an optimum training data selection strategy.

Finally, a KA iterative method is employed to ensure that actual target returns are excluded from the training data to avoid desensitization [55].

Figure 4.20 shows the results of applying the IKA-STAP architecture to the sixth KASSPER data challenge set [55]. Note that mainbeam and sidelobe clutter discretes have been eliminated, with a significantly reduced clutter residue overall and at the difficult land-sea interface (see [55] for further details regarding the underlying data set). Also contained in [55] are the results of applying IKA-STAP to recorded flight test data further establishing its efficacy.

Also developed under the auspices of the DARPA/AFRL KASSPER project was the GTRI KA-STAP (GKA-STAP) architecture shown in Figure 4.21, designed by researchers at the Georgia Tech Research Institute (GTRI) [7, 56]. While there are many complex elements to the overall architecture that we will not delve into in this forum, the reader is directed to the two-stage KA prefilter that effectively detrends the nonstationary aspects of the incoming multichannel data stream based on a number of knowledge sources as indicated (cultural database information, ownship position, calibration information, etc.) [57]. Figure 4.22 shows the corresponding performance of several variants of the GKA-STAP

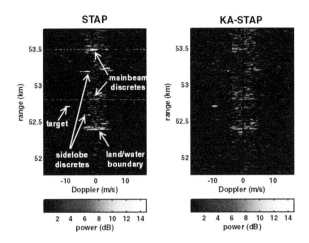

Figure 4.20 Application of the IKA-STAP architecture to the sixth KASSPER challenge data set [55]. Conventional STAP (left figure) results in significant clutter leakage and sidelobe discrete targets, while IKA-STAP essentially eliminates all of the complex clutter issues.

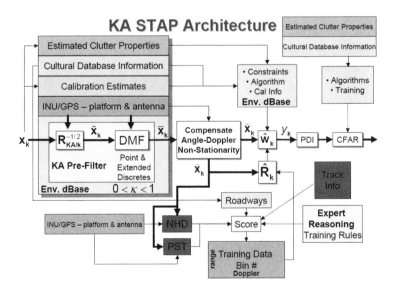

Figure 4.21 Illustration of the GTRI KA-STAP (GKA-STAP) architecture developed under the DARPA/AFRL KASSPER workshop. Though there a number of complex processing elements, the diagram clearly illustrates the central role that knowledge plays in detrending the nonstationary aspects of the incoming multichannel interference. Knowledge sources illustrated include cultural databases, ownship position information, and other calibration knowledge [57].

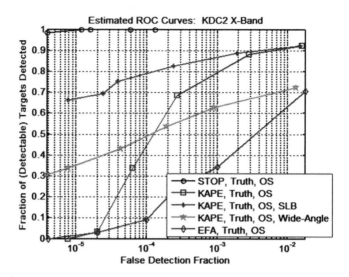

Figure 4.22 Estimated receiver operating characteristic (ROC) curves for the KASSPER challenge data set #2 for the clairvoyant (unobtainable) optimum (STOP) and several GTRI KA variants (prefix KAPE), as compared with a conventional post-Doppler STAP approach EFA. Note that all of the KA methods significantly outperform conventional STAP.

approach as compared with conventional STAP when applied to the KASSPER #2 challenge data set [57].

Yet another interesting KA architecture developed under the auspices of the KASSPER project utilized a physics-based approach to essentially estimate the parameters of all of the scatterers (targets and/or clutter) observable over some coherent processing interval [48]. Though very computationally intensive relative to conventional STAP (or the IKA or GKA architectures above), it is nonetheless extremely illuminating as it clearly shows the significant potential of a radar to precisely measure its environment.

Though the reader is referred to [48] for precise details, essentially the KA SCHISM (signal and clutter as highly independent structured odes) algorithm is fundamentally an estimation-subtraction approach similar to the CLEAN algorithm [58]—that is, as stronger modes are estimated, they are coherently subtracted from the CPI data cube. The process is then repeated ideally until the residue coincides with the radar noise floor (in practice this is only approximately achieved).

More precisely, over the coherent processing interval (CPI), M complex time samples are collected for each of N spatial channels.

Suppose the space and time samples are uniformly spaced Δx and Δt apart, respectively. The measured signal at the (n,m)-th space-time sample can be modeled as

$$y_{n,m}(\mathbf{A},\kappa,\omega) \equiv \sum_{p=1}^{p} A_p \exp\left[j(\kappa_p n + \omega_p m)\right] + v_{n,m} \qquad (4.21)$$

with

$$\kappa_p \equiv \frac{2\pi}{\lambda}\sin(\theta_p)\Delta_x \text{ and } \omega_p \equiv 2\pi f_p \Delta t_{n,m} \qquad (4.22)$$

where A_p is the complex amplitude of the pth structured (target or clutter) signal component, k_p is the spatial frequency associated with a plane wavefront arriving at a beam angle, θ_p, relative to broadside, ω_p is the temporal frequency associated with a Doppler frequency, f_p, and $v_{n,m}$ is the measurement noise. A point jammer is modeled with M or fewer strong Doppler frequencies from a single beam direction.

Application of the above results in a scatter map for all targets, clutter and/or moving targets is shown in Figure 4.23. Scatterers that are sufficiently separated from the angle-Doppler clutter ridge are declared moving targets. Also shown in Figure 4.23 are the results before and after array manifold calibration.

What is particularly intriguing about the SCHISM approach, is that it produces a high-resolution range-Doppler image that rivals traditional SAR imagery, but with a far shorter dwell time. Moreover, since it does not utilize a sample covariance matrix, clutter discretes are readily detectable. Consequently, even if SCHISM could not be implemented for every CPI in real-time, it could be used intermittently to populate the large clutter discrete portion of the EDDB database.

4.4 KA Radar Epilogue

A common misconception for those first encountering KA methods is that somehow only *a priori* knowledge is being used, such

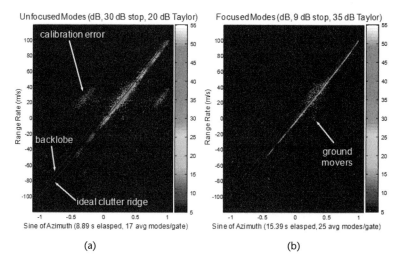

Figure 4.23 Sample outputs from physic-based KA SCHISM algorithm as applied to the original KASSPER challenge data set before (a) and after (b) array manifold calibration. By utilizing an underlying physics model, rather than a purely statistical sample covariance approach, individual scatterers (targets and clutter discretes) are readily discernable. Such data can in turn, be used to populate the EDDB.

as digital terrain databases [59]. Clearly this is not the case. Rather, a judicious blending of exogenous and endogenous data sources that draws on the relative strengths of each is the most effective. Both the IKA-STAP and GKA-STAP examples clearly illustrate this point. The key to KA processing is that knowledge, however it was obtained, is allowed to exist in a dynamical database (i.e., the EDDB) that can be exploited in real time. The look ahead HPEC architecture developed under the KASSPER project enables this approach.

Since the publication of the first edition of this book, KA radar methods have continued to be developed and refined. KA methods have been applied to a wide range of radar problems [10–23]. In some cases, entirely new radar applications have been enabled by KA methods. For example, the DARPA Multipath Exploitation Radar (MER) project successfully developed and demonstrated the ability for a GMTI radar to detect and track surface targets in an urban setting non-line-of-sight (NLOS) [14]. This latter application is the subject of a recent book [24].

In the next chapter we outline a new approach to cognitive radar, one that combines KA processing with *both* transmit and receive adaptivity to provide the most flexible and truly adaptive radar yet conceived.

References

[1] Guerci, J. R., and E. J. Baranoski, "Knowledge-Aided Adaptive Radar at DARPA: An Overview," *IEEE Signal Processing Magazine*, Vol. 23, 2006, pp. 41–50.

[2] Lawson, J. L., and G. E. Uhlenbeck, *Threshold Signals*, Vol. 24 of MIT Radiation Laboratory Series, New York: McGraw-Hill, 1950.

[3] Ridenour, L. N., *Radar System Engineering*, Vol. 1 of MIT Radiation Laboratory Series, New York: McGraw-Hill, 1947.

[4] Guerci, J. R., *Space-Time Adaptive Processing for Radar*, Norwood, MA: Artech House, 2003.

[5] Monzingo, R. A., and T. W. Miller, *Introduction to Adaptive Arrays*, SciTech Publishing, 2003.

[6] Ward, J., "Space-Time Adaptive Processing for Airborne Radar," *IEE Colloquium on Space-Time Adaptive Processing (Ref. No. 1998/241)*, 1998, p. 2.

[7] Melvin, W. L., and J. R. Guerci, "Knowledge-Aided Signal Processing: A New Paradigm for Radar and Other Advanced Sensors," *IEEE Transactions on Aerospace and Electronic Systems*, Vol. 42, 2006, pp. 983–996.

[8] Guerci, J. R., "Knowledge-Aided Sensor Signal Processing and Expert Reasoning (KASSPER)," *Proceedings of 1st Annual DARPA KASSPER Workshop*, Washington, D.C., 2002.

[9] Reed, I. S., J. D. Mallett, and L. E. Brennan, "Rapid Convergence Rate in Adaptive Arrays," *IEEE Transactions on Aerospace and Electronic Systems*, Vol. AES-10, 1974, pp. 853–863.

[10] Zhu, H., Z. Zhu, and F. Su, "Clutter Properties and Suppression Methods of Hyper Sonic Airborne Radar," in 2018 14th IEEE International Conference on Signal Processing (ICSP), 2018, pp. 859–862.

[11] Riedl, M., and L. C. Potter, "Knowledge-Aided Bayesian Space-Time Adaptive Processing," *IEEE Transactions on Aerospace and Electronic Systems*, Vol. 54, 2018, pp. 1850–1861.

[12] Riedl, M., and L. C. Potter, "Multimodel Shrinkage for Knowledge-Aided Space-Time Adaptive Processing," *IEEE Transactions on Aerospace and Electronic Systems*, Vol. 54, 2018, pp. 2601–2610.

[13] Yang, Z., and R. C. de Lamare, "Enhanced Knowledge-Aided Space-Time Adaptive Processing Exploiting Inaccurate Prior Knowledge of the Array Manifold," *Digital Signal Processing*, Vol. 60, 2017, pp. 262–276.

[14] Fertig, L. B., J. M. Baden, and J. R. Guerci, "Knowledge-Aided Processing for Multipath Exploitation Radar (MER)," *IEEE Aerospace and Electronic Systems Magazine*, Vol. 32, 2017, pp. 24–36.

[15] Bang, J., W. Melvin, and A. Lanterman, "Knowledge-Aided Covariance Matrix Estimation in Spiky Radar Clutter Environments," *Electronics*, Vol. 6, 2017, p. 20.

[16] New, D., and P. Corbell, "Interference Suppression Using Knowledge-Aided Subarray Pattern Synthesis," in *2016 IEEE Radar Conference (RadarConf)*, 2016, pp. 1–6.

[17] Kumbul, U., and H. T. Hayvaci, "Knowledge-Aided Adaptive Detection with Multipath Exploitation Radar," in *2016 Sensor Signal Processing for Defence (SSPD)*, 2016, pp. 1-4.

[18] Gao, Y., H. Li, and B. Himed, "Knowledge-Aided Range-Spread Target Detection for Distributed MIMO Radar in Nonhomogeneous Environments," *IEEE Transactions on Signal Processing*, Vol. 65, 2016, pp. 617–627.

[19] Wang, J., A. M. Ponsford, and W. Emily, "Knowledge Aided Detector," U.S. Patent 9,157,992 B2, obtained October 13, 2015.

[20] Riedl, M., and L. C. Potter, "Knowledge-Aided GMTI in a Bayesian Framework," in *2015 IEEE Radar Conference (RadarCon)*, 2015, pp. 1240–1243.

[21] Guerci, J., R. Guerci, A. Lackpour, and D. Moskowitz, "Joint Design and Operation of Shared Spectrum Access for Radar and Communications," in *2015 IEEE Radar Conference (RadarCon)*, 2015, pp. 0761–0766.

[22] Bang, J. H., W. L. Melvin, and A. D. Lanterman, "Model-Based Clutter Cancellation Based on Enhanced Knowledge-Aided Parametric Covariance Estimation," *IEEE Transactions on Aerospace and Electronic Systems*, Vol. 51, 2015, pp. 154–166.

[23] De Maio, A., A. Farina, and G. Foglia, "Knowledge-Aided Bayesian Radar Detectors and Their Application to Live Data," *IEEE Transactions on Aerospace and Electronic Systems*, Vol. 46, 2010, pp. 170–183.

[24] Watson, B., and J. R. Guerci, *Non-Line-of-Sight Radar*, Norwood, MA: Artech House, 2019.

[25] Capraro, C. T., et al., "Implementing Digital Terrain Data in Knowledge-Aided Space-Time Adaptive Processing," *IEEE Transactions on Aerospace and Electronic Systems*, Vol. 42, 2006, pp. 1080–1099.

[26] Capraro, C. T., et al., "Improved STAP Performance Using Knowledge-Aided Secondary Data Selection," *Radar Conference, Proceedings of the IEEE*, 2004, pp. 361–365.

[27] Capraro, G. T., et al., "Knowledge-Based Radar Signal and Data Processing: A Tutorial Review," *IEEE Signal Processing Magazine*, Vol. 23, 2006, pp. 18–29.

[28] Melvin, W., et al., "Knowledge-Based Space-Time Adaptive Processing for Airborne Early Warning Radar," *Aerospace and Electronic Systems Magazine, IEEE*, Vol. 13, 1998, pp. 37–42.

[29] Wicks, M. C., et al., "Space-Time Adaptive Processing: A Knowledge-Based Perspective for Airborne Radar," *IEEE Signal Processing Magazine*, Vol. 23, 2006, pp. 51–65.

[30] Hall, D., and J. Llinas, "An Introduction to Multisensor Data Fusion," *Proceedings of the IEEE*, Vol. 85, 1997, pp. 6–23.

[31] Nitzberg, R., *Radar Signal Processing and Adaptive Systems*. Norwood, MA: Artech House, 1999.

[32] Guerci, J. R., and W. Baldygo, *Proceedings of the DARPA/AFRL Knowledge-Aided Sensor Signal Processing and Expert Reasoning (KASSPER) Workshop*, 2002–2006.

[33] Gini, F., and M. Rangaswamy, "Knowledge-Based Radar Detection, Tracking and Classification," in *Adaptive & Learning Systems for Signal Processing, Communications & Control Series*, New York: Wiley-IEEE Press, 2008.

[34] Farina, A., and L. Timmoneri, "Real-time STAP techniques," *Electronics & Communication Engineering Journal*, Vol. 11, 1999, pp. 13–22.

[35] Klemm, R., *Principles of Space-Time Adaptive Processing* Vol. 12: Institution of Electrical Engineers, 2002.

[36] Guerci, J. R., J. S. Goldstein, and I. S. Reed, "Optimal and Adaptive Reduced-Rank STAP," *IEEE Transactions on Aerospace and Electronic Systems*, Vol. 36, 2000, pp. 647–663.

[37] Zatman, M., "Properties of Hung-Turner Projections and Their Relationship to the Eigencanceller," *Conference Record of the Thirtieth Asilomar Conference on Signals, Systems and Computers*, Vol. 2, 1996, pp. 1176–1180.

[38] Billingsley, J., *Low Angle Radarland Clutter: Measurements and Empirical Models*, William Andrew, 2002.

[39] Rangaswamy, M., J. Michels, and B. Himed, "Statistical Analysis of the Non-Homogeneity Detector for STAP Applications," *Digital Signal Processing*, Vol. 14, 2004, pp. 253–267.

[40] Goldstein, J., I. Reed, and L. Scharf, "A Multistage Representation of the Wiener Filter Based on Orthogonal Projections," *IEEE Transactions on Information Theory*, Vol. 44, 1998, pp. 2943–2959.

[41] Wishart, J., "The Generalized Product Moment Distribution in Samples from a Normal Multivariate Population," *Biometrika*, Vol. 20A, 1982, pp. 32–52, .

[42] Pillai, S., and C. Burns, *Array Signal Processing*, New York: Springer-Verlag, 1989.

[43] Hiemstra, J. D., and C. SAIC, "Colored Diagonal Loading," *Proceedings of the 2002 IEEE Radar Conference*, 2002, pp. 386–390.

[44] Bergin, J. S., et al., "STAP with Knowledge-Aided Data Pre-Whitening," *Radar Conference, Proceedings of the IEEE*, 2004, pp. 289–294.

[45] Guerci, J. R., "Theory and Application of Covariance Matrix Tapers for Robust Adaptive Beamforming," *Signal Processing, IEEE Transactions on [see also IEEE Transactions on Acoustics, Speech, and Signal Processing]*, Vol. 47, 1999, pp. 977–985.

[46] Gelb, A., *Applied Optimal Estimation*: MIT Press, 2002.

[47] Techau, P. M., et al., "Performance Bounds for Hot and Cold Clutter Mitigation," *Aerospace and Electronic Systems, IEEE Transactions on*, Vol. 35, 1999, pp. 1253–1265.

[48] Legters, G. R., and J. R. Guerci, "Physics-Based Airborne GMTI Radar Signal Processing," *Radar Conference, Proceedings of the IEEE*, 2004, pp. 283–288.

[49] Stoica, P., et al., "On Using A Priori Knowledge in Space-Time Adaptive Processing," *Signal Processing, IEEE Transactions on* [see also *Acoustics, Speech, and Signal Processing*], Vol. 56, 2008, pp. 2598–2602.

[50] Jao, J., A. Yegulalp, and S. Ayasli, "Unified Synthetic Aperture Space Time Adaptive Radar (USASTAR) Concept," MIT Lincoln Laboratory NTI-4, 2004.

[51] Horn, R. A., and C. R. Johnson, *Topics in Matrix Analysis*, Cambridge University Press, 1991.

[52] Farina, A., *Antenna-Based Signal Procesing for Radar Systems*, Norwood, MA: Artech House, 1992.

[53] Steinhardt, A., "ASAP: Where From Here?," *Proceedings of Adaptive Sensor Array Processing*, MIT Lincoln Laboratory, 2002.

[54] Skolnik, M. I., *Radar Handbook*, 3rd ed., New York: McGraw-Hill, 2008, pp. 44–45.

[55] Bergin, J. S., et al., "Evaluation of Knowledge-Aided STAP Using Experimental Data," *Aerospace Conference, IEEE*, 2007, pp. 1–13.

[56] Melvin, W. L., and G. A. Showman, "Performance Results for a Knowledge-Aided Clutter Mitigation Architecture," *Proceedings of the IET International Conference on Radar Systems*, Edinburgh, Scotland, 2007.

[57] Melvin, W. L., and G. A. Showman, "Knowledge-Aided STAP Architecure," *Proceedings of Knowledge-Aided Sensor Signal Processing and Expert Reasoning (KASSPER)*, Las Vegas, NV, 2004.

[58] Clark, B., "An Efficient Implementation of the Algorithm CLEAN," *Astronomy and Astrophysics*, Vol. 89, 1980, p. 377.

[59] "Performance Specification Digital Terrain Elevation Data (DTED)," National Geospatial Agency, May 23, 2000.

5

Putting it All Together: CoFAR

5.1 Cognitive Radar: The Fully Adaptive Knowledge-Aided Approach

Chapter 2 demonstrated the significant potential of adding transmit adaptivity in addition to conventional receiver adaptation (e.g., STAP). However, to realize joint transmit-receiver adaptation gains, an estimate of the composite radar channel is required (target, interference, clutter, etc.). As discussed in Chapters 3 and 4, such an estimate can often prove difficult to obtain due to a multitude of real-world effects that are often euphemistically referred to as nonstationary statistics. In practice, there may not be a sufficient set of i.i.d. training samples from which to extract high-quality information for adaptation.

Chapter 4 introduced KA methods and processing architectures to overcome many of the aforementioned real-world effects—particularly in the case of land clutter where both a priori and onboard generated databases either exist or can be constructed. The look-ahead (i.e., predictive or noncausal) computing architecture introduced in Chapter 4 facilitates the exploitation of the environmental dynamic database (EDDB) in real time.

What remains is simply to put it all together; that is, full adaptivity (both transmit *and* receive) enhanced with KA processing. The resulting fully adaptive KA radar can truly be argued to possess the essential elements of an intelligent or cognitive radar as described in Chapter 1: It possesses extensive knowledge of the environment and the reasoning ability to make appropriate decisions regarding transmit-receive adaptivity [1–3]. Hence, we have a cognitive fully adaptive radar [4–6].

5.1.1 A Cognitive Radar Architecture for GMTI

While there are limitless variants of a cognitive radar as described herein, we will outline a high-level architecture targeting a GMTI/SAR airborne radar with all of the essential elements of full adaptivity and KA processing.

Figure 5.1 depicts the essential elements of a GMTI/SAR CoFAR. At the core is a CoFAR real-time channel estimator (RTCE) that is supported by both a real-time KA environmental database as described in Chapter 4, and proactive transmit probing techniques to aid in estimating signal-dependent channel components (targets, clutter, and potentially jamming).

As depicted, the RTCE accepts inputs from a multitude of potential sources, including off-board data such as digital terrain and land cover/land use (LCLU) maps, as well as onboard generated data such as georegistered SAR images and existing tracking data. As described in Chapter 4, it is essential that the RTCE is in communication with the radar scheduler in order to perform lookahead processing of regions soon to be illuminated.

In contrast to the KA architecture of Chapter 4, we now see that the CoFAR coprocessor is in communication with an adaptive transmitter; that is, a transmitter that in general is capable of varying its spatiotemporal waveform modulation and possibly other DoFs such as polarization. This proactive aspect is indeed a unique feature of a cognitive radar and is reminiscent of the biological transmit adaptivity of a bat [1].

5.1 Cognitive Radar: The Fully Adaptive Knowledge-Aided Approach   161

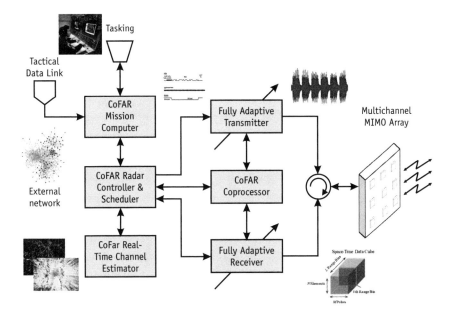

Figure 5.1 Example of a high-level CoFAR architecture that encompasses real-time channel estimation and full adaptivity (transmit and receive).

With knowledge of the transmit parameters, the adaptive receiver will thus in general perform KA adaptive pulse compression and STAP. While a multichannel monostatic configuration is shown in Figure 5.1, the above architecture is equally applicable to bi/multistatic configurations.

Architectural highlights include

- A RTCE updated with both off-board and onboard environmental data sources such as digital terrain and LCLU, georegistered SAR imagery, and GMTI tracks.
- A CoFAR coprocessor as described in Chapter 4 but with the additional function of providing adaptivity guidance to the transmitter as well as the receiver.
- An adaptive transmitter in addition to an adaptive receiver. The advent of digital front-ends, digital arbitrary waveform generators (DAWGs), and active electronically scanned arrays (AESAs) in modern radars has greatly facilitated this function.

- A CoFAR scheduler that is in communication with the RTCE and CoFAR coprocessor to provide the requisite look-ahead to compensate for database access latency (as described in Chapter 4). A discussion of the CoFAR scheduler is provided later in this chapter.

5.1.2 Informal Operational Narrative for a GMTI Radar

So how does a cognitive radar work—in plain English? To answer let's consider a typical ISR sortie that consists of surveilling a particular region, usually accomplished from a standoff racetrack type of orbit [7], as depicted in Figure 5.2. In general, the region of interest will be comprised of a multitude of different geographical and man-made features. Natural features might include land-sea interfaces, mountains, forests, and rivers, while man-made structures might include roads, power lines, towers, and urban centers. All of these features can cause abrupt changes in the underlying clutter statistics that, in turn, can cause degraded performance as described throughout this book and elsewhere [8–10]. Moreover,

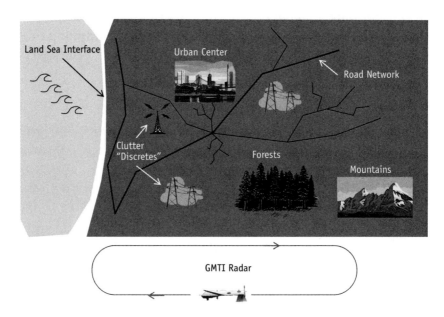

Figure 5.2 Illustration of a GMTI radar surveillance scenario depicting numerous real-world clutter effects that require a cognitive radar architecture.

many regions may be highly localized and thus not amenable to sample statistic-based covariance estimation.

In steady-state operation, the CoFAR coprocessor, in concert with the RTCE, is scanning ahead based on where the radar will shortly be illuminating per the radar scheduler. If, for example, a significant highway (and thus likely a significant number of vehicles) is present, the CoFAR coprocessor would determine how that road maps into radar coordinates (range, angle, and possibly Doppler) in order to segment the received data used for weights training (covariance estimation). For example, if the road is embedded in an otherwise reasonably homogeneous background (consider US 35 in remote parts of Texas), then the only action that the CoFAR coprocessor may invoke is to simply exclude those range bins containing highway traffic from the training set used for covariance estimation and adaptive CFAR thresholding [11, 12]. The same type of intelligent training data segmentation can be extended to a multitude of scenarios involving terrain transitions. Moreover, depending on the amount of available training data for a particular region, one can also optimize the ADoFs of the STAP filter (e.g., Example 4.1, Chapter 4).

For those regions where insufficient training data is available from the direct measurements, data associated with previous scans of that region and/or data synthesized from physical databases can be used to supplement the training data (as discussed in Chapter 4).

In populous areas, the RF spectrum may be crowded—resulting in potential spectral interference. As described in Example 2.1, it is possible to color the transmit spectrum through adaptive waveform modulation in order to antimatch to the interference. This is not only a good idea from an SINR perspective, but may be necessary to prevent interfering with others.

As demonstrated in Chapter 2, tailoring the transmit waveform can have a significant impact on target identification. Due to the potentially dense target background in many operational environments, this may be one of the most key aspects of adaptive modulation where there may be a strong desire to separate military from civilian targets.

When very large clutter discretes or targets are present in the sidelobes, it may also be desirable to prenull them by adapting the transmit spatial pattern as described in Chapter 2. In addition to reducing the impact of these interferers, it has the added benefit of better homogenizing the received data; that is, detrending out the spikey artifacts in the received data.

Finally, though not discussed in any detail in this book, KA tracking methods can be employed to make the most sense of the MTI detections [13]. Associating MTI detections with known road networks can increase final track accuracy; knowing when line-of-sight (LOS) dropouts will occur can allow for intelligent track coasting and/or association gate construction and so forth. Moreover, with the successful demonstration of NLOS detection and tracking of road vehicles in an urban setting under the DARPA MER project [14, 15], this will allow for even more sophisticated KA tracking.

As previously described, the RTCE is indeed dynamic in the sense that it too is continually updating its knowledge of the environment. For example, previously unknown road networks can often be discovered by GMTI radars. Through proper data quality vetting, updated database entries will be included to ensure proper handling of those regions on subsequent revisits. Also, as most GMTI radars host a SAR mode, radar imagery is processed for features of interest (large discretes, regions of inhomogeneity, etc.) and used to further update/enhance the RTCE.

5.2 CoFAR Radar Scheduler

Implicit throughout this book is the integrated availability of a real-time, automated intelligent decision-making process—a major assumption! In real-world challenging environments, a radar will be confronted with many *competing* requirements. For example, in the previous GMTI ISR radar case, there will arise conflicts between allocating valuable radar resources and timeline to *both* WAS *and* HVT tracking. Indeed, it is this fundamental conflict that in no small part led to the development and proliferation of AESAs that have the ability to rapidly and electronically switch transmit/receive beams, thus minimizing regret when conducting simultaneous track and search (see [16] for an excellent introduction to AESAs).

The CoFAR radar scheduler performs several major functions:

- Creates a RTCE model derived from *all* of the aforementioned mechanisms from Chapters 2 through 4; namely, active MIMO probing and KA channel estimation.
- Performs a real-time resource allocation optimization that attempts to divvy up the finite resources (energy, time, computation, etc.) to best meet both mission level requirements and evolving tactical requirements (e.g., electronic protection, emissions control (EMCON)).
- Finally, a transmit timeline is created down to each transmit pulse (and possibly intrapulse), the time-horizon of which can vary widely from a fraction of a second to many minutes.

Figure 5.3 shows a high-level lowdown from the main aircraft mission computer (in this case, ISR mission computer) to the CoFAR radar scheduler. The scheduler, in turn, performs all functions necessary to populate the complete radar timeline—down to potentially sub-PRI time frames.

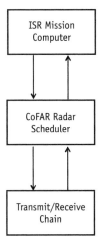

Figure 5.3 High-level illustration of the flow-down from the mission computer to the CoFAR scheduler.

Mathematically, the scheduler receives real-time (and generally continuous) inputs from the mission computer and crafts an objective function J that both quantifies overall performance relative to mission objectives, and applies hard and soft constraints. An apt metaphor for the CoFAR scheduler is the conductor of an orchestra.

As an example, consider the following example ISR objective function:

$$J(\mathbf{x}) = J_{WAS}(\mathbf{x}) + J_{HVT}(\mathbf{x}) + J_{SAR}(\mathbf{x}) + J_{CR}(\mathbf{x}) \qquad (5.1)$$

where

$$\begin{aligned}
J_{WAS}(\mathbf{x}) &= \text{Wide-area search metric} \\
J_{HVT}(\mathbf{x}) &= \text{High-value target track metric} \\
J_{SAR}(\mathbf{x}) &= \text{Synthetic aperture radar metric} \\
J_{CR}(\mathbf{x}) &= \text{Cognitive radar functions (e.g., MIMO probing)}
\end{aligned} \qquad (5.2)$$

The above would, of course, be subject to a multitude of constraints on a variety of factors, including, for example, transmit duty factor, peak power, spatial and/or frequency keep-out zones.

Optimal radar and cognitive radar scheduler design is an entire discipline onto itself. The reader is referred to [17] for further details.

5.3 Areas for Future Research and Development

This book is most decidedly an introduction. Cognitive radar is still an area of active research. The following is a list of areas requiring further development (in no particular order):

1. Further development of KA algorithms that mature the concepts of intelligent filter selection and training and fine-grained KA covariance/statistical estimation. This is perhaps the easiest instantiation of a cognitive radar.
2. Further development and maturation of the EDDB concept, including new knowledge sources, vetting of database en-

tries to ensure requisite accuracies, and KA methods designed to operate with imperfect database entries.
3. Further development and maturation of the look-ahead scheduling approach to optimize overall cognitive radar efficiency (i.e., performance per unit processor resources). This includes methods for establishing KA processing duty factor or load balancing; that is, methods for determining which regions require KA processing given finite computational, memory, database resources, and so forth.
4. Continued development and maturation of adaptive/KA transmitter techniques including adaptive waveform diversity and spatiotemporal beamforming.
5. Further development of a CoFAR scheduler for a variety of radar applications.

References

[1] Haykin, S., "Cognitive Radar: A Way of the Future," *IEEE Signal Processing Magazine*, Vol. 23, 2006, pp. 30–40.

[2] Guerci, J. R., "Next Generation Intelligent Radar," in *2007 IEEE Radar Conference*, 2007, pp. 7–10.

[3] Guerci, J. R., M. C. Wicks, J. S. Bergin, P. M. Techau, and S. U. Pillai, "Theory and Application of Optimum and Adaptive MIMO Radar," in *2008 IEEE Radar Conference*, IEEE, 2008, pp. 1-6.

[4] Zasada, D. M., J. J. Santapietro, and L. D. Tromp, "Implementation of a Cognitive Radar Perception/Action Cycle," in *IEEE Radar Conference*, Cincinnati, OH, 2014, pp. 0544–0547.

[5] Guerci, J. R., R. M. Guerci, M. Ranagaswamy, J. S. Bergin, and M. C. Wicks, "CoFAR: Cognitive Fully Adaptive Radar," in *IEEE Radar Conference*, Cincinnati, OH, 2014, pp. 0984–0989.

[6] Guerci, J. R., "Cognitive Radar: The Next Radar Wave?" *Microwave Journal*, Vol. 54, January 2011, pp. 22–36.

[7] Dunn, R. J., P. T. Bingham, and C. A. Fowler, *Ground Moving Target Indicator Radar and the Transformation of US Warfighting*, www.grumman.com/analysis-center/images/pdf/Ground-Moving-Target-Indicator-gmti-brief.pdf, 2004.

[8] Melvin, W. L., and J. R. Guerci, "Knowledge-Aided Signal Processing: A New Paradigm for Radar and Other Advanced Sensors," *IEEE Transactions on Aerospace and Electronic Systems*, Vol. 42, 2006, pp. 983–996.

[9] Guerci, J. R., and E. J. Baranoski, "Knowledge-Aided Adaptive Radar at DARPA: An Overview," *IEEE Signal Processing Magazine,* Vol. 23, 2006, pp. 41–50.

[10] Guerci, J. R., "Knowledge-Aided Sensor Signal Processing and Expert Reasoning (KASSPER)," presented at the Proceedings of 1st Annual DARPA KASSPER Workshop, Washington, DC, 2002.

[11] Nitzberg, R., *Radar Signal Processing and Adaptive Systems,* Norwood, MA: Artech House, 1999.

[12] Melvin, W., M. Wicks, P. Antonik, Y. Salama, L. Ping, and H. Schuman, "Knowledge-Based Space-Time Adaptive Processing for Airborne Early Warning Radar," *IEEE Aerospace and Electronic Systems Magazine,* Vol. 13, 1998, pp. 37–42.

[13] Gini, F., and M. Rangaswamy (eds.), *Knowledge Based Radar Detection, Tracking and Classification,* New York: Wiley-IEEE Press, 2008.

[14] Fertig, L. B., J. M. Baden, and J. R. Guerci, "Knowledge-Aided Processing For Multipath Exploitation Radar (MER)," *IEEE Aerospace and Electronic Systems Magazine,* Vol. 32, 2017, pp. 24–36.

[15] Watson, B. C., and J. R. Guerci, *Non-Line-of-Sight Radar,* Norwood, MA: Artech House, 2019.

[16] Sturdivant, R., C. Quan, and E. Chang, *Systems Engineering of Phased Arrays,* Norwood, MA: Artech House, 2018.

[17] Nguyen, H. K., *Optimal Radar Scheduler Design.* Norwood, MA: Artech House, in preparation.

6

Cognitive Radar and Artificial Intelligence

6.1 Relationship between Cognitive Radar and Artificial Intelligence

At the time of the writing of this second edition, there remains a degree of ambiguity in the terms cognitive systems and artificial intelligence (AI) and/or machine intelligence/learning. To be clear, cognitive radar is a subset of AI. Machine learning more recently referred to as deep learning is also a subset of AI, albeit a more recent addition. As described in Chapter 1 and repeated below, a cognitive radar can be succinctly described as an automated system that has *sophisticated* real-time environmental/contextual knowledge that can be used to effect *advanced* real-time adaptivity of both transmit and receive functions. Note the emphasis on "sophisticated" and "advanced." An amoeba has rudimentary environmental awareness, but you would not use it as a meaningful exemplar of a cognitive system. A circa 1940s analog CFAR circuit is an example of rudimentary real-time adaptivity, but clearly not a good example of "advanced" adaptivity. The degree of sophistication in environmental awareness *and* real-time adaptivity is what warrants the new moniker of "cognitive." The original example of the bat proposed by Haykin remains a good original biological analog for cognitive active sensing [1].

It is absolutely possible to build and deploy a fully cognitive radar as described herein without ever using any of the so-called modern machine learning techniques such as deep learning (e.g., convolutional neural networks [2]). Of course, it is also possible to develop a cognitive radar that fully leverages the latest advances in AI. We will briefly discuss some of the possible applications later in this chapter. Figure 6.1 depicts the basic relationship between AI and cognitive radar. Note that KA methods, as described in Chapter 4, draw on the more traditional AI methods (e.g., expert systems).

6.2 Cognitive Radar Utilizing Traditional AI

In this section, we will revisit the example CR architecture of Chapter 5, first using more traditional AI methods (see Figure 6.1), then in Section 6.3 using modern deep-learning AI.

Referring to Figure 5.1, we see a depiction of the essential elements of a GMTI/SAR CoFAR. As previously described, the

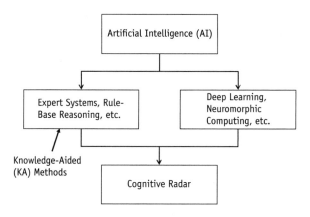

Figure 6.1 Relationship between modern AI and cognitive radar.

CoFAR RTCE is supported by both a real-time KA environmental database as described in Chapter 4, and proactive transmit probing techniques to aid in estimating signal-dependent channel components (targets, clutter, and potentially jamming). In an expert systems (ES) instantiation, both the interpretation of results/measurements and the decision process are aided by ES AI methods. For example, in a rules-based ES, a measurement result "large abrupt increase in ground clutter level measured occurring near land-sea interface in terrain database" would be interpreted as "land-sea interface identified." Note how a measurement (clutter level change) was compared to a knowledge database (terrain database) to provide confirmation and confidence in the decision process. Additional rules-based methods could also be employed to further minimize the probability of false positives (e.g., abrupt *and* sustained increase in ground clutter levels further indicative of a land-sea interface).

ESs, as the name implies, attempt to mimic the behavior of a human expert or subject matter expert (SME). Thus, continuing with the above example, "What would a SME do with the knowledge of a land-sea interface?" One obvious action is to separate clutter returns into two regions: land and sea, applying different CFAR and STAP algorithms to each. More refined questions such as "What sea state?" and "What terrain type?" could also be asked by the ES to further refine the actions to be undertaken.

In the first edition of this book published in 2010, the emphasis was on KA methods (i.e., more traditional AI methods, such as the above example). In the next section, we explore areas where the newer deep learning AI could be applied to CR.

6.3 Cognitive Radar Utilizing Deep Learning AI

It is hard to overstate the impact that recent advances in deep learning (DL) have had in numerous applications from image processing to world-class championship gaming [3, 4]. While there are many efforts underway to apply DL to many diverse problem sets, currently, the most readily applicable (and successful) problems are pattern recognition in nature; that is, recognizing (identifying) objects in an image or equivalent abstracted observation space. This immediately suggests one major area in cognitive radar where DL should be directly applicable: target ID.

One major area in radar target ID is SAR. In the 1990s, DARPA and AFRL undertook a major project titled The Moving and Stationary Target Acquisition and Recognition (MSTAR) [5]. From [5], "The goal of this project was to advance the state of Automatic Target Recognition (ATR) using synthetic aperture radar (SAR) imagery by developing the technology of model-based vision" A comprehensive data repository can be found in [6]. The database consists of 1-ft (square pixel) resolution SAR images at X-band. Some sample images are shown in Figure 6.2. Recent examples where DL has been successfully applied to radar applications such as this can be found in [7–9].

A note of caution: ground targets remain notoriously difficult for ATR due to a variety of real-world factors including aspect

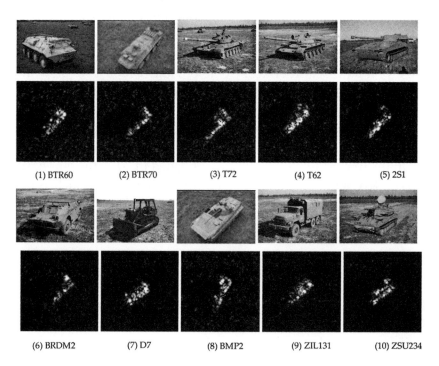

Figure 6.2 Sample DARPA MSTAR SAR X-band SAR images of various military ground vehicles (optical images top rows, grayscale SAR images bottom rows); see [6]. SAR target ID is one major are where modern AI techniques are directly applicable.

6.3 Cognitive Radar Utilizing Deep Learning AI

angle dependence, partial obscuration, and/or within-class configuration variations. At the time of this writing, there still is no fully automated ATR algorithm that is truly robust to the aforementioned deleterious factors. Airborne targets are, not surprisingly, a bit easier to tackle since (aside from crabbing) are generally pointed in the direction they are flying, are not obscured, and are not embedded in background clutter (for the most part).

There are of course many other areas where DL could be applied in a CoFAR architecture. Referring to Figure 6.3, we will step through how modern AI might be able to be applied to different CoFAR subsystems.

6.3.1 CoFAR Mission Computer

At the end of the day, warfare is a game—albeit a potentially deadly one. It thus stands to reason that gaming applications of modern DL AI *could* be adapted to the CoFAR mission computer [10]. This would particularly be the case as the "game" unfolds. However, a

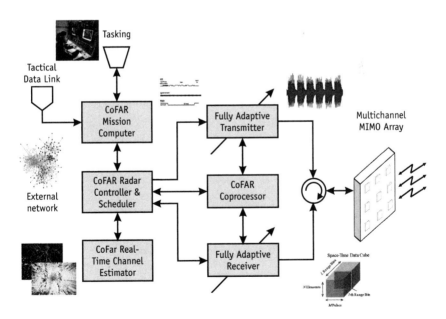

Figure 6.3 CoFAR architecture has many opportunities for the insertion of modern AI techniques.

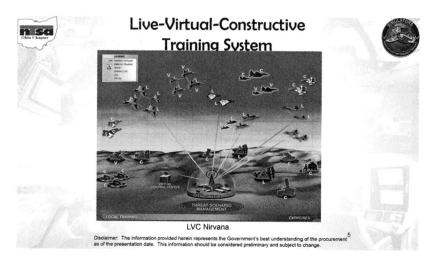

Figure 6.4 The new LVC training environment envisioned by the US DoD.

key prerequisite would be to create a highly realistic gaming environment in which the DL could train. At the time of this writing, a major initiative is underway by the United States Department of Defense to create a highly realistic synthetic warfare gaming environment dubbed the Live Virtual Construct (LVC) (see Figure 6.4 and [11]). While focused on human-in-the-loop (HIL) training, it is conceivable that this could be adopted for AI; that is LVC-AI. Of course, to be effective, the gaming environment would have to include enormous variability in adversarial and environmental factors—sources of major uncertainty in the real world. One major advantage of AI compared to humans is the ability to tirelessly execute millions of "what-if" sorties. This means that an AI-trained mission computer could be more effective than a human—in exactly the same way that DeepMind is a better Go player than the world's champion human Go player! [10].

6.3.2 CoFAR Radar Controller and Scheduler

As the name implies, the CoFAR radar controller and scheduler translates overarching mission requirements from the mission computer (MC), along with inputs from the RTCE, to establish the radar mode sequencing and transmit-receive timeline. In general, this is a highly complex, highly nonlinear data assignment problem as described in Chapter 5. As such, modern AI techniques that are focused on the solution of highly nonlinear, generally nonconvex, multidimensional optimization problems *might* be applicable [12]. Although again, enormous amounts of training data and/or a robust training environment would be required to ensure robustness and reliability when applied to real-world applications.

6.3.3 CoFAR RTCE

It is true to say: "As goes channel knowledge, so goes performance!" Therefore, anything and everything that can be utilized to maximize channel knowledge is fair game. Indeed, the very need for cognitive radar is based on a need for sophisticated estimation and adaptation to an ever-changing complex radar channel. As this book makes clear, there are many ways to enhance channel knowledge, from a priori KA methods as developed under the DARPA KASSPER project [13], to proactive MIMO probing techniques as described in Chapter 3. Not surprisingly, there is virtually no limit to where modern AI methods could be injected into the CoFAR RTCE. AI pattern recognition techniques could be used to effect trouble recognition techniques when scanning large digital terrain/environmental databases. AI could also be applied to the multitude of decision theory and optimization problems that are ubiquitous in the CoFAR RTCE process. Again, there really is no limit.

6.4 Summary

At the time of this writing, there has been a veritable explosion in the development of new advanced AI technique-based DL. This chapter described how cognitive radar fits into the overall scheme of AI in general, and how new AI techniques might be applied to a number of CoFAR functions. However, there are many radar problems that still present major challenges even to the most advanced AI techniques. For an additional cautionary perspective, the reader is referred to the paper by Greenspan [14].

References

[1] Haykin, S., "Cognitive Radar: A Way of the Future," *IEEE Signal Processing Magazine*, vol. 23, pp. 30-40, 2006.

[2] Krizhevsky, A., I. Sutskever, and G. E. Hinton, "Imagenet Classification with Deep Convolutional Neural Networks," in *Advances in Neural Information Processing Systems*, 2012, pp. 1097–1105.

[3] Silver, D., A. Huang, C. J. Maddison. et al., "Mastering the Game of Go with Deep Neural Networks and Tree Search," *Nature*, Vol. 529, 2016, p. 484.

[4] LeCun, Y., Y. Bengio, and G. Hinton, "Deep Learning," *Nature*, Vol. 521, 2015, pp. 436–444.

[5] Hummel, R., "Model-Based ATR Using Synthetic Aperture Radar," in *Record of the IEEE 2000 International Radar Conference [Cat. No. 00CH37037]*, 2000, pp. 856–861.

[6] *MSTAR Data Repository*, https://www.sdms.afrl.af.mil/index.php?collection=mstar.

[7] Mason, E., B. Yonel, and B. Yazici, "Deep Learning for Radar," in *2017 IEEE Radar Conference (RadarConf)*, 2017, pp. 1703–1708.

[8] Schwegmann, C. P., W. Kleynhans, B. P. Salmon, L. W. Mdakane, and R. G. Meyer, "Very Deep Learning for Ship Discrimination in Synthetic Aperture Radar Imagery," in *2016 IEEE International Geoscience and Remote Sensing Symposium (IGARSS)*, 2016, pp. 104–107.

[9] Gong, M., J. Zhao, J. Liu, Q. Miao, and L. Jiao, "Change Detection in Synthetic Aperture Radar Images Based on Deep Neural Networks," *IEEE Transactions on Neural Networks and Learning Systems*, Vol. 27, 2015, pp. 125–138.

[10] Gibney, E., "Google AI Algorithm Masters Ancient Game of Go," *Nature News*, Vol. 529, 2016, p. 445.

[11] *Live Virtual Constructive (LVC)*, https://www.ndia.org/-/media/sites/ndia/meetings-and-events/5180-ntsa/2017/71a0/briefings-2017/live-virtual-construction_lvc.ashx.

[12] Adler J., and O. Öktem, "Solving Ill-Posed Inverse Problems Using Iterative Deep Neural Networks," *Inverse Problems*, Vol. 33, 2017, p. 124007.

[13] Guerci, J. R., and E. J. Baranoski, "Knowledge-Aided Adaptive Radar at DARPA: An Overview," *IEEE Signal Processing Magazine*, Vol. 23, 2006, pp. 41–50.

[14] Greenspan, M., "Potential Pitfalls of Cognitive Radars," in *2014 IEEE Radar Conference*, 2014, pp. 1288–1290.

About the Author

Joseph R. Guerci has over 30 years of advanced technology development experience in industrial, academic, and government settings—the latter included a seven-year term with Defense Advanced Research Projects Agency (DARPA), where he led major new radar developments. The author of over 100 technical papers and publications, he is a fellow of the IEEE for Contributions to Advanced Radar Theory and its Embodiment in Real-World Systems, the recipient of the IEEE Warren D. White Award for Excellence in Radar Adaptive Processing and Waveform Diversity, and the recipient of IEEE Dennis J. Picard Medal for Contributions to Advanced, Fully Adaptive Radar Systems and Real-Time Knowledge-Aided, and Cognitive Radar Processing Architectures.

Index

A

Active electronically scanned arrays (AESAs), 40, 161, 164–65
Adapted space-time transmit pattern, 96
Adapted transmit pattern, 115–16
Adaptive beamforming
   SINR loss performance for, 90
   on transmit, 96
Adaptive degrees-of-freedom (ADoFs)
   defined, 127
   matching to training data, 128–35
   spatial and temporal, 130
   twice the number rule, 130
Adaptive MIMO radar
   dynamic calibration, 91–93
   introduction to, 87–88
   nonorthogonal MIMO probing for channel estimation, 106–16
   optimization problem, 88
   theoretical performance bounds of DDMA MIMO STAP approach, 101–6
   transmit-dependent channel estimation, 93–101
   transmit-independent channel estimation, 88–91
Adaptive multipath interference mitigation, 90–91
Adaptive receivers, 19, 24, 161
Adaptive transmit capability, 19–24
Adaptive transmitter
   channel knowledge and, 24
   feedback from receive chain and, 19

Adaptive transmitter (continued)
    further development of, 167
Adaptivity, 19
Additive colored Gaussian noise
    (AGCN) case
    about, 38
        maximizing output SINR for, 41
        multichannel radar block diagram
            for, 39
        optimum receiver for, 40
        Additive Gaussian white noise
            (AGWN), 112
Additive noise jamming, 88–89
Airborne MTI radar
    back lobe radiation, 55
    clutter suppression, 54–60
    STAP-Tx example, 94–97
Air Force Research Laboratory
    (AFRL), 124, 125
Angle-Doppler clutter, 55
Angle-Doppler coupling
    coefficient, 56
Angle-Doppler point targets, 95
Angle-Doppler transmit
    pattern, 59, 60
Area coverage rate (ACR), 31
Artificial intelligence (AI)
    cognitive radar and, 12, 169–75
    deep learning, cognitive radar
        utilizing, 171
    relationship between cognitive
        radar and, 169–70
    traditional, cognitive radar
        utilizing, 170–71
Automated gain control (AGC), 119
Automatic Target Recognition
    (ATR), 172, 173

B

Bayesian approach, 131–34

Binary hypothesis testing
    problem, 60–61

C

Calibration, dynamic MIMO, 91–93
Cauchy-Schwarz theorem, 43
Cell averaging constant false alarm
    rate (CACFAR), 119
Channel estimation
    nonorthogonal MIMO probing for,
        106–16
    transmit-dependent, 93–101
    transmit-independent, 88–91
Channel kernel matrix, 76
CLEAN algorithm, 134, 152
Clutter
    angle-Doppler, 55
    distributed, 112
    homogeneous, 124
    inhomogeneity, 120
    iso-range ring, 54
    kernel matrix, 57
    reflectivity random variable of, 52
    suppression, 54–60
    uncorrelated, 57
    uniformly illuminated, 95
    uniform random, 53
Clutter dominant case, 49
Clutter transfer matrix, 49, 51, 94
Cochannel narrowband interferers, 48
Code division multiple access
    (CDMA), 96–97
CoFAR scheduler
    about, 164–65
    artificial intelligence and, 174–75
    communication with RTCE and
        coprocessor, 162
    flow-down to, 165
    functions, 165
    further development of, 167
    real-time inputs, 166
Cognitive fully adaptive radar
    (CoFAR)

artificial intelligence and, 173–75
conventional modern radar versus, 16, 18–19
coprocessor, 160, 161, 163
high-level block diagram, 18
mission computer, 173–74
radar controller, 174–75
real-time channel estimator (RTCE), 160, 161, 164, 165, 175
See also CoFAR scheduler
Cognitive radar (CR)
architecture, 16–32
artificial intelligence and, 12, 169–75
benefits of, 16
confusion and debate about, 11
conventional modern radar versus, 16, 18–19
fully adaptive knowledge-aided approach, 159–64
functional elements and characteristics of, 16–32
future research and development, 166–67
for GMTI, 160–62
mapping of biological cognitive properties of, 16
optimum resource allocation, 29–32
popularity and interest in, 11
resource optimization, 19
scheduling, 29–32
Coherent change detection (CCD) methods, 135
Coherent pulse interval (CPI), 94–97, 138, 152
Colored loading, 147, 148
Colored noise
impact on shaping transmit probe, 45
interference, 44

interference, spectra of, 46
multipath and, 45
Constant false alarm rate (CFAR), 124–26
Constant modulus
method of stationary phase and, 72–74
NLFM to achieve, 74–76
Constrained optimization, 71
Constrained optimum MIMO radar
constant modulus and method of stationary phase, 72–76
GMSP approach, 76–80
linear constraints case, 67–70
nonlinear constraints case, 70–72
prenulling on transmit, 68–70
recent advances in, 80–82
relaxed projection approach, 70–72
Conventional modern radar, 16, 18
Covariance matrix
colored noise interference, 44–45
from decorrelating process, 136
DoF MIMO STAP, 101
total interference, 41
Covariance matrix tapering (CMT), 135–37
Covariance training data selection strategy, 122

D

DARPA
Mountain Top radar, 122
MSTAR, 172
Multipath Exploitation Radar (MER) project, 154
See also KASSPER project
Data prewhitening, 134–35
DDMA MIMO STAP clutter mitigation
about, 97–99
covariance matrix, 101

DDMA MIMO STAP clutter mitigation (continued)
  illustrated, 100
  MIMO STAP Doppler steering vector, 103
  operating sequence, 99–101
  resolution recovery, 106
  signal flow diagram, 100
  theoretical performance bounds of, 101–6
  three-element STAP array versus, 105
Deep learning (DL), 12, 171–73
Degrees-of-freedom (DoFs)
  adaptive (ADoFs), 127, 128–35
  number of transmit and receive, 40
  spatio-temporal, 120
  target identification and, 20
  temporal, in CPI, 122
  transmit, 97–99, 103, 104
Deterministic covariance, 134
Deterministic point target, 58
Digital arbitrary waveform generators (DAWGS), 19, 161
Digital terrain and elevation data (DTED), 126
Direct KA radar, 131–35
Distributed clutter, 109–10, 112
Doppler division multiple access (DDMA)
  defined, 97
  MIMO clutter mitigation, 97–101
  See also DDMA MIMO STAP clutter mitigation
Doppler filtering, 97
Duty factor, 167
Dynamic MIMO calibration, 91–93

E

Electronic protection emissions control (EMCON), 165

Environmental database manipulation, 143
Environmental dynamic database (EDDB)
  defined, 19
  element tags, 144–45
  further development of, 166–67
  integration of, 24
  large discretes and, 146
  memory-based knowledge, 27

F

Fat-tails, 135
FIR filter, 44
Fully adaptive radar (FAR), 107

G

Generalized matched subspace projection (GMSP) approach
  example, 80
  matched subspace, 76
  matched subspace calculation, 79
  matched subspace projection, 77–78
  minimum SINR gain requirement, 79
  multichannel transmit input optimization determination, 78–79
  projecting nominal waveforms, 80
  sequential search, 80
  steps, 78–80
  transmit criteria setting, 79
GKA-STAP
  about, 150
  architecture illustration, 151
  defined, 29
  variants of, 150–52
Gram-Schmit formulation, 27
Green's function
  CNR versus, 114, 115

composite, 113
defined, 107
ideal fast-time clutter comparison, 108
linear nature of, 110
MIMO channel, 110
spatial, 113
Ground moving target indicator (GMTI) radar
   cognitive radar architecture for, 160–62
   DDMA MIMO clutter mitigation for, 97–101
   distributed clutter, 109–10
   FDMA-like Doppler frequency modulation for, 103
   IKA-STAP archtitecture and, 148
   information operational narrative for, 162–64
   ISR, 164
   MDV and, 96
   real-world environmental effects of, 16, 17
   road networks and, 121
   SAR mode, 164
   slow movers and, 123
   surveillance scenario, 162

H

High-performance embedded computing (HPEC), 126
High-value targets (HVTs), 91
Hyperspectral imagery, 126–27

I

IKA-STAP
   about, 146–47
   application of, 30, 151
   architecture illustration, 147
   colored loading covariance estimate, 147
   defined, 29
   GMTI application, 148
   land cover-based training data editing, 149
   optimum clutter discrete handling, 147
   processed snapshot vector, 149
   result of applying, 150, 151
   subsystems, 149, 150
Indirect KA radar, 126–28
Inertial navigation system (INS), 26
Infinite duration (steady state) case, 84–85
Input-output transfer, 44
Interference statistics estimation, 89
Internal clutter motion (ICM), 120, 131

J

Jamming
   additive noise, 88–89
   transmit configuration and, 48

K

KA algorithms, 166
KA covariance, 148
KA-embedded computing architecture, 27, 125
KA-HPEC architecture
   balancing throughput in, 143
   defined, 140
   example, 142
   high-level canonical, 144
   key breakthrough in, 140–41
Kalman filter, 28, 133
KA radar
   Bayesian approach to, 131–34
   data prewhitening, 134–35
   direct, 131–35
   dynamic environmental data and, 129
   epilogue, 153–55

KA radar (continued)
  fully adaptive, 159–64
  indirect, 126–28
  introduction to, 124–39
  past observations as prior
    knowledge source, 135–39
  real-time, 139–53
  throughput constraint, 145
KA SCHISM, 152–53
KASSPER project
  DARPA and AFRL partnership
    and, 125
  estimated ROC curves for, 30
  GKA-STAP, 150–52
  goal of, 27
  IKA-STAP, 146–50, 151
  KA architectures developed by,
    146–53
  key KA-HPEC breakthrough in,
    140–41
  look-ahead techniques, 28
  96-node real-time, 141, 142
  SCHISM, 152–53
  workshops, 146
KA-STAP
  architecture illustration, 25
  direct and indirect
    methods, 126–28
  EDDB as component of, 19
  GKA-STAP, 29, 150–52
  high-level canonical HPEC
    architecture, 144
  IKA-STAP, 29, 30, 146–50, 151
  pipelining throughput to, 143
KB/KA algorithms, 125
Keep-out constraints, 68–69
Knowledge-aided (KA) processing
  coprocessor, 25
  development of, 24
  dynamic environmental data
    and, 129
  full adaptivity and, 160
  key to, 154
  origins of, 26
  performance improvement
    examples, 26
  space-time clutter characteristics
    and, 123
  special, 25
Knowledge-based space-time
  adaptive processing (KB-
  STAP), 26, 27, 125
Kronecker delta function, 102
Kronecker product, 136

L

Land cover land use (LCLU), 126, 160
Linear constraints case, 67–70
Linear frequency modulation (LFM)
  long-pulse case and, 46
  pulse spectra comparison, 73
  short-pulse case and, 46
  unoptimized, 72
Linear time invariant (LTI) case, 39
Linear whitening filter, 42
Load balancing, 167

M

Machine learning. *See* Artificial
  intelligence (AI)
Matched subspace, 76, 79
Matched subspace projection, 77–78
Maximum likelihood estimation
  (MLE), 121, 132
Method of stationary phase, 74–76
MIMO cohere-on-target, 91–93
MIMO radar
  adaptive, 87–116
  adaptive beamforming on transmit
    and, 96
  block diagram, 39
  case, 108–16
  clutter mitigation, 97–101

constrained optimum, 67–80
DDMA, 20
knowledge-aided, 59
optimum relations, 37
optimum target
    identification, 60–67
orthogonal techniques, pros and
    cons, 109
spatial, 108
transmit-receive
    optimization, 38–48
MIMO STAP Doppler steering
    vector, 103
Minimum detectable velocity
    (MDV), 96, 123
Monte Carlo trials, 90
Moving and Stationary Target
    Acquisition and Recognition
    (MSTAR), 172
MTI radar
    airborne, 54–60, 94–97
    side-looking, 128
Multidimensional echo, 38
Multielement arrays, 94
Multipath Exploitation Radar (MER)
    project, 154
Multipath interference, 43
Multipath transfer matrix, 44–45
Multiple input, multiple output. *See*
    MIMO radar
Multistate Weiner filter (MWF), 127
Multitarget case
    about, 64–66
    example, 66–67
    optimizing transmit input and, 67
    target impulse responses, 66, 67

N

Noise
    additive, 88–89
    AGWN, 112
    colored, 44, 45, 46
    filtering in presence of, 111
    white, 44–45
    whitening, 42
Nonhomogeneity detector
    approach (NHD), 129
Nonlinear constraints case, 70–72
Nonlinear FM
    to achieve constant modulus, 74–76
    constant modulus
        waveform, 74, 75
    method of stationary phase and, 73
    waveforms, 73, 75–76
Nonorthogonal MIMO (NO-MIMO)
    probing, 106–16
    probing beam illustration, 114
    waveforms, 106
Nonzero norm requirement, 111
Normalized angle, 50

O

Optimal pulse shape
    for detecting a point target, 53
    for maximizing SCR, 51–53
Optimum MIMO (OptiMIMO), 107
Optimum space-time MIMO
    processing, 54–60
Organization, this book, 11–13, 32
Over-the-horizon (OTH) radar
    modes, 107

P

Parseval's theorem, 74
Point targets
    angle-Doppler, 95
    deterministic, 58
    optimal pulse shape for
        detecting, 53
    unity gain, 45
Prenulling
    clutter edge, 95

Prenulling (continued)
   large discretes, 112
   on transmit, 68–70
Proactive sidelobe target blanking, 50
Pulse repetition frequency (PRF), 97
Pulse shape, 51–53

Q

QR-factorization, 27, 148

R

Rayleigh quotient, 49
Real-time channel estimator (RTCE)
   artificial intelligence and, 175
   in CoFAR architecture, 161
   CoFAR scheduler and, 162
   defined, 160
   as dynamic, 164
   model, 165
Real-time KA radar, 139–53
Receivers
   adaptive, 24, 161
   adaptivity, 19
   higher-order STAP, 97
   STAP, 56
   structure for binary hypothesis testing problem, 61
Reed, Mallet, Brennan (RMB) rule, 137
Reflectivity, estimated, 133
Relaxed projection approach
   about, 70–71
   example, 71–72
   illustrated, 72
   waveform with LFM-like properties and, 72
Resource allocation
   conflicts, 31
   importance of, 29
   optimum, 29–32

S

Sample matrix inverse (SMI) approach, 128
Schedulers, 28
Schwarz's inequality, 84
Sense-learn-adapt (SLA)
   channel probing techniques, 12
   perception cycle, 31, 32
Sidelobe target
   proactive blanking, 50
   suppression, 49–51
Signal-to-clutter ratio (SCR)
   level, setting, 58
   maximizing, 38, 51, 58
   at receiver input, 48–49
Signal-to-interference-plus-noise ratio (SINR)
   for adaptive beamforming, 90
   for long-pulse case, 47
   loss, output, 90
   maximizing, 20, 38–48
   multipath interference and, 43–48
   optimal pulse shape for maximizing, 51
   optimum pulse, 48
   for short-pulse case, 47
   STAP-Tx method, 98
Single-input, single-output (SISO), 22, 107, 108
Space-time adaptive processing (STAP)
   beamformer, 27
   filter, 56
   filter design, 120
   receiver performance of, 56
   reduced-risk methods, 130
   space-time transmit patterns and, 54
   See also KA-STAP
Space-time clutter transfer matrix, 94

Space-time MIMO processing, 54–60
Space-time snapshot vector, 134
Space-time steering vector, 133
Stale weights problem, 133
STAP-Tx method
  about, 94–95
  adapted pattern, 96
  example, 94–97
  form, 95–96
  orthogonal waveforms, 96–97
  performance potential, 96
  SINR performance, 98
Subject matter expert (SME), 171
Synthetic aperture radar (SAR), 126

T

Tapped delay line filter, 44
Target identification
  DoFs and, 20
  optimum MIMO, 60
  transmit optimization and, 23
Target transfer matrix, 39, 45
Time-division multiple access
    (TDMA), 109, 112
"Training" period, 24
Transmit
  adapted pattern, 115–16
  adapted space-time pattern, 96
  adaptive beamforming on, 96
  adaptive capability, 19–24
  DoFs, 97–99, 103, 104
  jamming and, 48
  optimization, example, 23
  optimizing jointly with
      receive, 48–60
  optimum waveforms, 21, 47
  prenulling on, 68–70
  reconstruction, 103–4
  spatial pattern, tailoring, 22
  steering vectors, 107, 111
Transmit adaptation algorithms, 19

Transmit antenna pattern nulls, 51
Transmit-dependent channel
    estimation
  about, 93–94
  channel characteristics, 93
  DDMA MIMO STAP clutter
      mitigation example, 97–101
  STAP-Tx example, 94–97
Transmit independent channel
    estimation
  about, 88
  adaptive multipath interference
      mitigation, 90–91
  additive, 88–89
  interference statistics estimation, 89
Transmitters
  adaptive, 19, 161, 167
  degrees-of-freedom (DoFs), 19
Two-target ID problem
  example of, 63–64
  illustrated, 61
  maximum separation, 62–63
  optimum pulse spectrum, 65
  pulse modulation, 64
  SNR level, 63–64
  target impulse responses, 63
  transmit waveforms, 64

U

Uniform linear array (ULA), 113
Unity gain point target, 45

W

Whitening filter, 45
Whitening properties, 41–42
White noise, 44, 45
Wide area search (WAS) mode, 31
Wishart distribution, 131–32

X

X-band radar measurements, 124

Recent Titles in the Artech House Radar Series

Dr. Joseph R. Guerci, Series Editor

Adaptive Antennas and Phased Arrays for Radar and Communications, Alan J. Fenn

Advanced Techniques for Digital Receivers, Phillip E. Pace

Advances in Direction-of-Arrival Estimation, Sathish Chandran, editor

Airborne Pulsed Doppler Radar, Second Edition, Guy V. Morris and Linda Harkness, editors

Basic Radar Analysis, Mervin C. Budge, Jr. and Shawn R. German

Basic Radar Tracking, Mervin C. Budge, Jr. and Shawn R. German

Bayesian Multiple Target Tracking, Second Edition, Lawrence D. Stone, Roy L. Streit, Thomas L. Corwin, and Kristine L Bell

Beyond the Kalman Filter: Particle Filters for Tracking Applications, Branko Ristic, Sanjeev Arulampalam, and Neil Gordon

Cognitive Radar: The Knowledge-Aided Fully Adaptive Approach, Second Edition, Joseph R. Guerci

Computer Simulation of Aerial Target Radar Scattering, Recognition, Detection, and Tracking, Yakov D. Shirman, editor

Control Engineering in Development Projects, Olis Rubin

Design and Analysis of Modern Tracking Systems, Samuel Blackman and Robert Popoli

Detecting and Classifying Low Probability of Intercept Radar, Second Edition, Phillip E. Pace

Digital Techniques for Wideband Receivers, Second Edition, James Tsui

Electronic Intelligence: The Analysis of Radar Signals, Second Edition, Richard G. Wiley

Electronic Warfare in the Information Age, D. Curtis Schleher

Electronic Warfare Target Location Methods, Second Edition, Richard A. Poisel

ELINT: The Interception and Analysis of Radar Signals, Richard G. Wiley

EW 101: A First Course in Electronic Warfare, David Adamy

EW 102: A Second Course in Electronic Warfare, David Adamy

EW 103: Tactical Battlefield Communications Electronic Warfare,
David Adamy

FMCW Radar Design, M. Jankiraman

Fourier Transforms in Radar and Signal Processing, Second Edition, David Brandwood

Fundamentals of Electronic Warfare, Sergei A. Vakin, Lev N. Shustov, and Robert H. Dunwell

Fundamentals of Short-Range FM Radar, Igor V. Komarov and Sergey M. Smolskiy

Handbook of Computer Simulation in Radio Engineering, Communications, and Radar, Sergey A. Leonov and Alexander I. Leonov

High-Resolution Radar, Second Edition, Donald R. Wehner

Highly Integrated Low-Power Radars, Sergio Saponara, Maria Greco, Egidio Ragonese, Giuseppe Palmisano, and Bruno Neri

Introduction to Electronic Defense Systems, Second Edition, Filippo Neri

Introduction to Electronic Warfare, D. Curtis Schleher

Introduction to Electronic Warfare Modeling and Simulation, David L. Adamy

Introduction to RF Equipment and System Design, Pekka Eskelinen

Introduction to Modern EW Systems, Andrea De Martino

An Introduction to Passive Radar, Hugh D. Griffiths and Christopher J. Baker

Introduction to Radar using Python and MATLAB®, Lee Andrew Harrison

Linear Systems and Signals: A Primer, JC Olivier

Meter-Wave Synthetic Aperture Radar for Concealed Object Detection, Hans Hellsten

The Micro-Doppler Effect in Radar, Second Edition, Victor C. Chen

Microwave Radar: Imaging and Advanced Concepts, Roger J. Sullivan

Millimeter-Wave Radar Targets and Clutter, Gennadiy P. Kulemin

MIMO Radar: Theory and Application, Jamie Bergin and Joseph R. Guerci

Modern Radar Systems, Second Edition, Hamish Meikle

Modern Radar System Analysis, David K. Barton

Modern Radar System Analysis Software and User's Manual, Version 3.0, David K. Barton

Monopulse Principles and Techniques, Second Edition, Samuel M. Sherman and David K. Barton

MTI and Pulsed Doppler Radar with MATLAB®, Second Edition,
  D. Curtis Schleher

Multitarget-Multisensor Tracking: Applications and Advances Volume III,
  Yaakov Bar-Shalom and William Dale Blair, editors

Non-Line-of-Sight Radar, Brian C. Watson and Joseph R. Guerci

Precision FMCW Short-Range Radar for Industrial Applications,
  Boris A. Atayants, Viacheslav M. Davydochkin, Victor V. Ezerskiy,
  Valery S. Parshin, and Sergey M. Smolskiy

Principles of High-Resolution Radar, August W. Rihaczek

Principles of Radar and Sonar Signal Processing,
  François Le Chevalier

Radar Cross Section, Second Edition, Eugene F. Knott, et al.

Radar Equations for Modern Radar, David K. Barton

Radar Evaluation Handbook, David K. Barton, et al.

Radar Meteorology, Henri Sauvageot

Radar Reflectivity of Land and Sea, Third Edition, Maurice W. Long

Radar Resolution and Complex-Image Analysis, August W. Rihaczek and
  Stephen J. Hershkowitz

Radar RF Circuit Design, Nickolas Kingsley and J. R. Guerci

Radar Signal Processing and Adaptive Systems, Ramon Nitzberg

Radar System Analysis, Design, and Simulation, Eyung W. Kang

Radar System Analysis and Modeling, David K. Barton

Radar System Performance Modeling, Second Edition, G. Richard Curry

Radar Technology Encyclopedia, David K. Barton and
  Sergey A. Leonov, editors

Radio Wave Propagation Fundamentals, Artem Saakian

Range-Doppler Radar Imaging and Motion Compensation,
  Jae Sok Son, et al.

Robotic Navigation and Mapping with Radar, Martin Adams,
  John Mullane, Ebi Jose, and Ba-Ngu Vo

Signal Detection and Estimation, Second Edition, Mourad Barkat

Signal Processing in Noise Waveform Radar, Krzysztof Kulpa

Signal Processing for Passive Bistatic Radar, Mateusz Malanowski

Space-Time Adaptive Processing for Radar, Second Edition,
  Joseph R. Guerci

Special Design Topics in Digital Wideband Receivers, James Tsui

Systems Engineering of Phased Arrays, Rick Sturdivant, Clifton Quan, and
  Enson Chang

Theory and Practice of Radar Target Identification, August W. Rihaczek and Stephen J. Hershkowitz

Time-Frequency Signal Analysis with Applications, Ljubiša Stanković, Miloš Daković, and Thayananthan Thayaparan

Time-Frequency Transforms for Radar Imaging and Signal Analysis, Victor C. Chen and Hao Ling

Transmit Receive Modules for Radar and Communication Systems, Rick Sturdivant and Mike Harris

For further information on these and other Artech House titles, including previously considered out-of-print books now available through our In-Print-Forever® (IPF®) program, contact:

Artech House
685 Canton Street
Norwood, MA 02062
Phone: 781-769-9750
Fax: 781-769-6334
e-mail: artech@artechhouse.com

Artech House
16 Sussex Street
London SW1V HRW UK
Phone: +44 (0)20 7596-8750
Fax: +44 (0)20 7630-0166
e-mail: artech-uk@artechhouse.com

Find us on the World Wide Web at: www.artechhouse.com